THE PHYSIOLOGY OF PHYSICAL TRAINING

THE PHYSIOLOGY OF PHYSICAL TRAINING

ZSOLT RADÁK

ACADEMIC PRESS

An imprint of Elsevier

Academic Press is an imprint of Elsevier
125 London Wall, London EC2Y 5AS, United Kingdom
525 B Street, Suite 1800, San Diego, CA 92101-4495, United States
50 Hampshire Street, 5th Floor, Cambridge, MA 02139, United States
The Boulevard, Langford Lane, Kidlington, Oxford OX5 1GB, United Kingdom

Library of Congress Cataloging-in-Publication Data
A catalog record for this book is available from the Library of Congress

British Library Cataloguing-in-Publication Data
A catalogue record for this book is available from the British Library

ISBN: 978-0-12-815137-2

For information on all Academic Press publications
visit our website at https://www.elsevier.com/books-and-journals

Publishing Editor: Mica Haley
Acquisition Editor: Mary Preap
Editorial Project Manager: Tracy Tufaga
Production Project Manager: Punithavathy Govindaradjane
Cover Designer: Matthew Limbert

Typeset by SPi Global, India

This book is dedicated
to my mother,
and to Gabi, Fanni, Bence, and Hanna

CONTENTS

LIST OF FIGURES AND TABLES

ABOUT THE AUTHOR

Zsolt Radák received his PhD from the University of Tsukuba in Japan. He has received Hungary's Bolyai Research Fellowship in Medicine and the Szechenyi Professorship from the Hungarian Academy of Science. He is the DSc of the Hungarian Academy of Science in the field of medicine. He is a professor and served as dean, and also associate-dean, of the Faculty of Physical Education and Sport Science of Semmelweis University, Budapest, Hungary. Since 2008 he has been the head of the only sport science doctoral school in Hungary. From September of 2014 he has been a professor and head of the Research Institute of Sport Science, University of Physical Education. He has been a visiting professor at Toho University, Juntendo University, Waseda University, Japan, Texas University, United States, National Taiwan Sport University, Taiwan, and Beijing Sport University, China. He edited the book of *Free Radicals in Exercise and Aging*, published by Human Kinetics, and *Exercise and Diseases*, published by Meyer & Meyer. He published more than 200 papers in academic journals.

PREFACE

Physical performance has been in vogue for centuries. Physical training was an essential part of everyday activities of athletes preparing for the ancient Olympic Games. The influential effect of diet on physical performance has been recognized for more than a thousand years. The starting point of science-based physical training appears to have arisen in an exercise physiology book published in the United States in 1860. The establishment of the first academic unit in exercise physiology at Harvard University in 1891 provided a firm base to amalgamate science and sport performance. August Sternberg Krogh, a Danish physiologist, was awarded the Nobel Prize in 1920 for the discovery of the mechanism of regulation of the capillaries in skeletal muscle. Archibald Vivian Hill, of Cambridge University, was awarded the Nobel Prize in 1922 for the elucidation of mechanical work in muscles. Scientific discoveries in the 20th century had a huge positive impact on physical performance. The enormous differences between the results of the ancient Olympics and the modern Olympics are the result of well-designed training programs which are based on scientific discoveries. Indeed, scientific research is an essential part of the design of modern physical training.

Sport science focuses on exercise, outstanding physical performance, and health. From the middle of twentieth century daily physical work has gradually disappeared from everyday life due to technical development; it has been replaced by physical inactivity in the developed world. Since physical activity was an influential factor in genetic development of the human body, physical inactivity has disturbed the balance between lifestyle and genetic characteristics. Lifestyle diseases have emerged and science, including sport science, has shed light on the role that physical exercise can play in health prevention. Early work was reported to have been done on prevention by Leonardo da Vinci, who noticed, by dissecting human tissues, that degeneration of blood vessels affected aristocrats more than peasants. He hypothesized, investigated, and recorded his observations, and part of his work (approximately 7000 pages) have been preserved. He recognized the preventive effects of physical training in the 16th century, and modern scientific research confirmed his findings in the 1970s. Today it is clear that physical inactivity increases the risk of the development of lifestyle diseases, and that physical training is essential to live a healthy life. It is also clear that physical inactivity decreases average life expectancy.

ACKNOWLEDGMENTS

I was very lucky during my time in Japan because I learned from excellent professors such as Katsumi Asano and Hideki Ohno. I am so pleased that I can say that my postdoctoral supervisor Sataro Goto is now a friend of mine. Later on, I was lucky again because I met with Prof. Mitsuru Higuchi, who is my great supporter and also my great friend. I would like to express my sincere thanks for the long-term support of my friends, Jozsef Dubecz, Prof. Jozsef Tihanyi, Prof. Zoltan Benyo, and my Canadian friend Prof. Albert W. Taylor. The artistic images in this book are down to the excellent work of Dr. Erika Koltai. I would also like to acknowledge the fantastic work of Elsevier staff: Stacy Massuci, Tracy Tufaga, and Punithavathy Govindaradjane.

INTRODUCTION

The development of conditioning abilities is essential in every sport. These abilities influence vitality, quality of life, and the capacity to prevent diseases in addition to sport performance. This book covers the physiological and methodological aspects of physical training, which provide essential knowledge for everyone involved in sport. Physiological processes at the cellular level and the whole organism are discussed to better explain a particular training method, which provides the opportunity to convey deeper knowledge and understanding. Appreciation, integration, and implementation of scientific discoveries will be essential for outstanding sport performance in international competitions in the future. This book combines sport physiology with knowledge about the molecular mechanisms behind training methods. Adaptation responses are discussed in detail as one of the most important and complex elements in enhancing performance. Conditioning abilities are important in everyday life as well as sport. Thus, in this book we discuss prevention and aging. The frequency and quality of everyday physical activity of the elderly declines when compared to those of young and middle-aged populations; therefore, frequent physical exercise will have a significant positive impact on their lives.

At the end of each chapter the reader will find a series of test questions, intended to check the knowledge gained. A bibliography provides a list of additional reading. An "R Extra" section at the end of each chapter contains additional information for those who wish to read advanced material; this is more applicable to graduate students.

To comprehend the material presented in this book requires a working knowledge of exercise physiology and methodology, since the basic fundamentals are not explained in detail here. Only a part of the Dubecz pie chart (Fig. I.1) is explained from a physiological perspective. The purpose of this book is to integrate recent discoveries of sport science into practice and to stimulate students to pursue the latest information on the effects of exercise training on sport performance and conditioning techniques, and the effects of regular physical activity on health promotion.

It must be noted that the contents of his book introduce only a small part of recent scientific discoveries in sport science, and practical implementations that have led to outstanding scientific performance and health prevention.

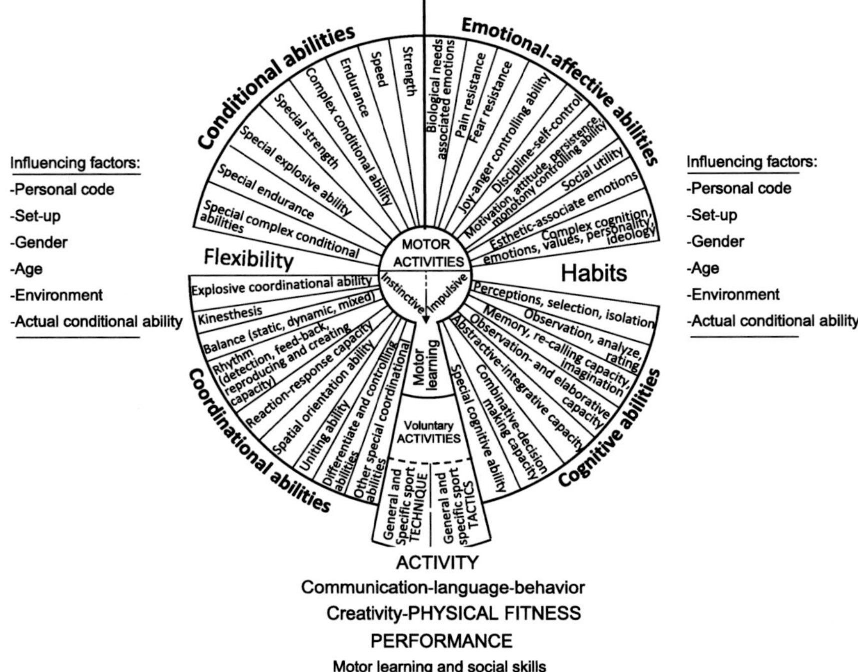

Fig. I.1 Dubecz pie chart. The effect of physical training on personality is very complex.

CHAPTER 1

Basic Cellular Functions, Cellular Adaptation, and Metabolism

The cell is the basic unit of all living organisms containing cell organelles and limited by its cell membrane. In higher developed organisms, cells are specialized and cooperate with each other to a great extent. The success in evolutionary selection depends on the viable information of DNA and the ability of the cells to adapt to environmental factors.

The discovery of DNA and decoding of its information was one of the greatest discoveries in the 20th century, and lets us explore and further understand the functions of the human body, heredity of phenotypical traits, and mechanisms of diseases; we can also use decoded genetic information to determine the role of genetic variations in sport performance and competences. Knowledge of genetic factors in sport performance is a basic requirement in individual sport selection. In this chapter, we address cellular functions since a thorough knowledge of basic cell biology is essential to understand more complex physiological and methodological processes and functions, which enable us to influence these processes to our advantage within limits.

1.1 CELL AND ORGANELLES

Cell cytoplasm is surrounded by a cell membrane; in the case of a muscle cell or fibers, this is called sarcolemma. The cell membrane is composed of a double layer of phospholipid molecules which are partly hydrophobic facing inside and partly hydrophilic facing outside. Embedded within this membrane is a variety of protein molecules that act as channels and pumps, transport proteins that move different molecules into and out of the cell, and structural proteins. Membrane receptors are specified according to their structure and function. For example, growth factor receptors, after binding to their hormone, initiate a process that stimulates protein synthesis, which results in an increase in size in the case of a muscle. Insulin receptor activation leads to the translocation of GLUT4 transport protein to the cell

The Physiology of Physical Training
https://doi.org/10.1016/B978-0-12-815137-2.00001-2

membrane, which allows the uptake of sugar molecules from the blood by the musculature.

One of the most important roles of cell membrane is its involvement in the transport processes of the cell, which can be active (not requiring energy) or passive (driven by a concentration gradient). A concentration gradient is required for the stimulation of a cell. This gradient is maintained by pumps; a Na–K pump works against the ion gradient by using the energy of ATP molecules, and pumps out Na^+-ions and pumps in K^+-ions. Inside the cell in the cytoplasm, there are several organelles, which are in close association with the membrane (Fig. 1.1).

The endoplasmic reticulum (EPR) plays an important role in cell protein and lipid transport, and in protein synthesis in conjunction with ribosomes, which are associated to the cell's membrane. EPR in muscle is known as sarcoplasmic reticulum, which functions as Ca^{++} ion storage. In response to stimulation, Ca^{++} ions flood the sarcoplasm, and upon binding to troponin they allow the formation of actomyosin complex, resulting in muscle contraction.

Active reuptake of Ca^{++} ion to the sarcoplasmic reticulum is required for muscle relaxation, which is catalyzed by a Ca^{++}-ATPase enzyme. Thus, Ca^{++} ions are necessary for muscle contraction; however, a high intracellular Ca^{++} concentration inhibits muscle relaxation, causing contracture.

DNA is located in the nuclear chromatin organized by histone proteins, which protect the integrity of the genetic information. DNA is vulnerable

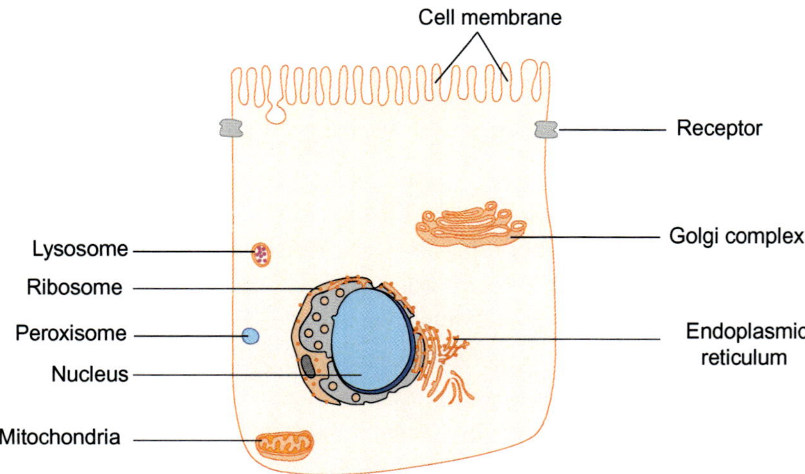

Fig. 1.1 Cell basic structure. A simplified graphic show the basic cell components and organelles.

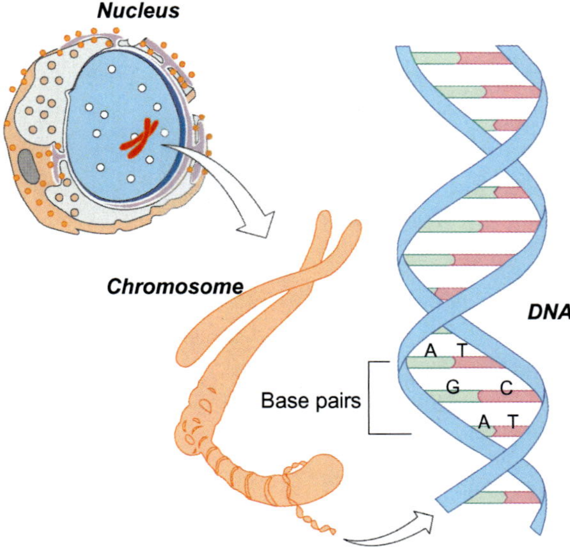

Fig. 1.2 A chromosome. Nuclear DNA stores genetic information specific to individuals, which are encoded by nucleotide base pairs.

because of its huge size (uncoiled length 1.7–8.5 cm), so it is crucial to preserve its structure against mechanical and chemical impacts. Nuclear DNA encodes genetic information using 3 billion nucleotides (Fig. 1.2).

DNA is organized into chromosomes, and an organism's complete set of DNA is called a genome. In the nucleus of every human somatic cell there are 46 chromosomes (made up of 23 pairs), which in turn contain $50 \times 10^6 - 250 \times 10^9$ base pairs, which are compressed in a $1.9\,\mu m$ area in a compact structure.

The capacity of human DNA would allow the encoding of 3 million average size proteins; however, only a portion of the whole DNA encodes proteins. A gene is a region (locus) in DNA that encodes information of proteins in the form of base pairs. The main function of the genome is to store the information for gene transcription and consequential protein expression. Upon stimulation of the cell (e.g., hormone-ligand binding in a cell membrane), gene transcription may be initiated, leading to mRNA synthesis and consequential protein synthesis (e.g., motor proteins in skeletal muscle). It is worth noting that 90% of DNA does not encode proteins, and the function of the silent parts is not fully understood, nor its influence on the remaining 10%.

DNA contains segments called pseudogenes, similar to genes, but its transcription does not lead to protein synthesis, or even inhibit it. The functions of pseudogenes are not yet well understood; however, their number and variations show differences in a higher volume in individuals and also between species than those present in genes. Therefore further examination of pseudogenes may have a more profound effect than the study of genes in sport performance and individual sport selection in the future.

DNA can be found in mitochondria as well. Mitochondria have a size of 0.5–12 μm, and are surrounded by a double membrane. Mitochondrial DNA (mtDNA) encodes proteins of oxidative phosphorylation, and other proteins of mitochondria are synthesized in the nucleus and transported to the mitochondria. In the inner membrane, electrons generated from nutrients flow through the electron transport chain, and are translocated to molecular oxygen resulting in H_2O molecules, and energy in the form of ATP used by cells (Fig. 1.3).

Skeletal muscle and cardiac muscle contain large amounts of mitochondria in the proximity of myofibrils, by which most of the ATP is metabolized.

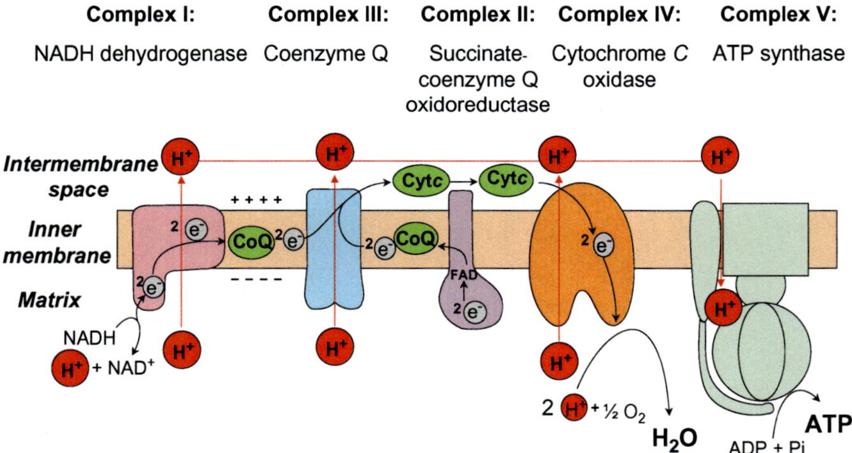

Fig. 1.3 Mitochondrial electron transport chain. Electron transport chain in the inner membrane of mitochondria comprises five protein complexes. Complex I pumps out protons to the space between the inner and outer membrane of the mitochondria using NADH as an energy source, and transfers electrons to Complex III. Complex II is a parallel electron transfer pathway to Complex I transferring electrons from FADH2 to Complex III, but unlike Complex I, no protons are transported to the intermembrane space. Complex III transfers electrons to Complex IV and pumps four protons per electron. Complex IV pumps two protons into the intermembrane space before electrons are transferred to molecular oxygen, leading to the production of the final metabolite, H_2O. Complex V is an ATP synthase, which acts as an ion channel that provides for a proton flux back into the mitochondrial matrix. This reflux releases energy, which is used to drive ATP synthesis.

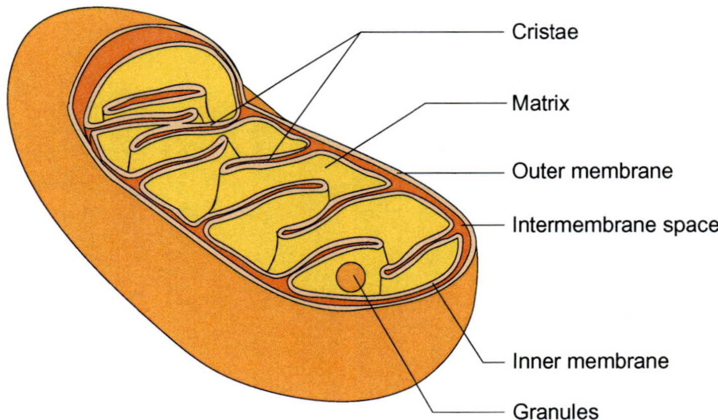

Fig. 1.4 Mitochondria. Mitochondria are an ATP-producing unit of the organism, and also play a role in programmed cell death and adaptation following physical training.

Mitochondrial membranes mainly influence the characteristics of these cell components. Most of the mitochondrial enzymes are located inside the mitochondria, in the matrix. These enzymes play a role in pyruvate, lipid, and Krebs-cycle (see later). The complexes embedded in the inner membrane produce the ATP from pyruvate, which is derived from glycolysis or lipid oxidation. Lipid metabolism is more efficient compared to carbohydrates since lipids contain six times more energy than the same amount of glycogen. All the glycogen stored in the cells can be burned in 1 day, whereas lipid storage is able to provide energy for a month. Thus, mitochondria play an important role in energy-producing processes and ATP production (Fig. 1.4).

In skeletal muscle, subsarcolemmal (below the membrane) and intermyofibrillar (close proximity to myofibrils) mitochondria are distinguished based more on their locations in the cell and less on their function. The number of mitochondria can be elevated significantly by physical training. This will be addressed in detail in Chapter 2.

Another important organelle of the cell is lysosome containing digestive enzymes, which break down proteins. These enzymes play a role in balancing the metabolism and catabolism of proteins. This balance is an important part of cellular homeostasis.

1.2 CELLULAR ADAPTATION

Upon stimulation, the cell responds to the stimulus with changes in protein synthesis. On the cell surface there are many sensors embedded in the membrane. Upon stimulation, two types of responses are possible. If a

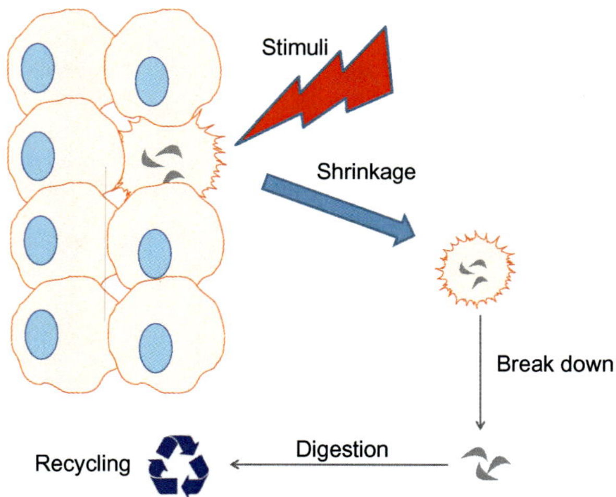

Fig. 1.5 Programmed cell death. An extremely strong stimulus may cause such damage to the cell that it would be transferred to its progenies, which would induce the development of a disease. Cells defend against such an event; therefore they commit suicide through programmed cell death.

stimulus is extremely strong, and the cell cannot maintain its function, it induces a signal, which leads to the synthesis of proteins involved in programmed cell death, called apoptosis. This process results in cell shrinkage and degradation of nuclear chromatin. The remaining degraded molecules can be used for biological processes. Apoptosis is a part of the defense system of the cell—for example, cells containing damaged or mutated DNA are eliminated through apoptosis, reducing the chance of the development of cancer (Fig. 1.5).

Another adaptation process beyond apoptosis to respond to stimuli is the production of proteins that increase the fitness of the cell (Fig. 1.6).

Defending against energy depletion cells increases the number of mitochondria, so the production of ATP increases; as a result, the cell is able to maintain its balance. Upon activation, the stimulus activates messenger molecules, which are phosphorylated, acetylated, or methylated, and then the signal is transferred into the nucleus, where transcription factors bind to DNA. Transcription factors are proteins that enter the nucleus upon stimuli and bind to certain sequences of DNA; therefore DNA double helix unseals, resulting in gene transcription and mRNA synthesis.

These sections encode proteins that will respond to stimuli (Fig. 1.7). Upon DNA transcription, DNA unseals and is transcribed to mRNA, which are comprised of coding called exon, and noncoding silent regions

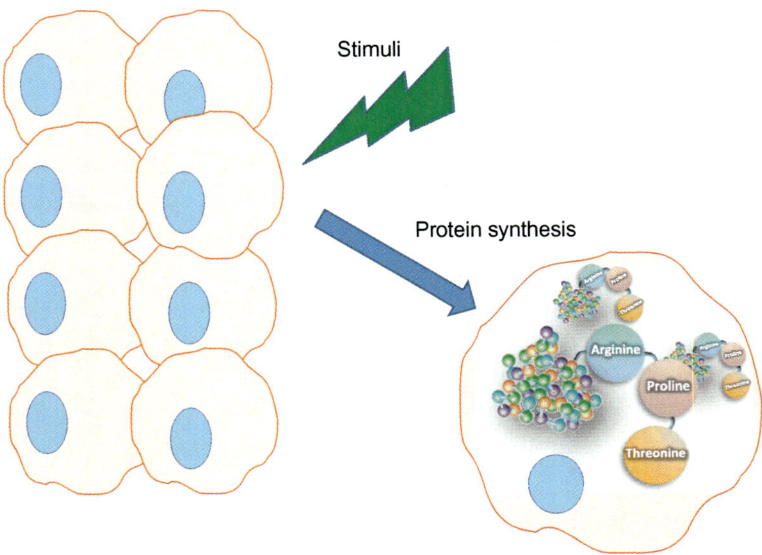

Fig. 1.6 Cellular adaptation. Cellular adaptation results from changes in protein synthesis which produces hypertrophy, an increase in the number of mitochondria, or an increase of vascular density.

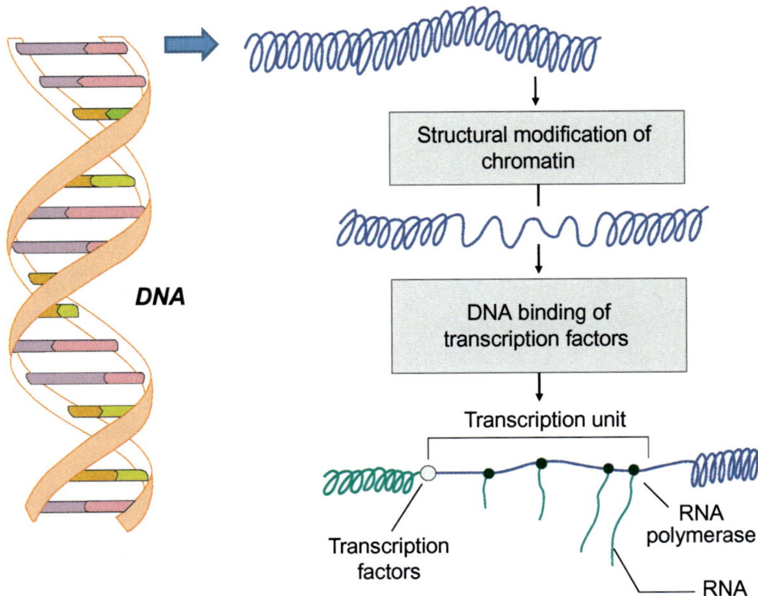

Fig. 1.7 Steps of transcription. An illustration of the steps of gene transcription, the first part of proteins' expression.

called intron. In the cytosol, mRNA is transcribed to proteins on the surface of ribosomes. The newly synthesized proteins then take their place in the cell and participate in distinct cellular processes, resulting in increased cell vitality.

Lifespans of proteins are different; some proteins have short lifetimes (perhaps hours), and are activated more quickly upon stimuli than proteins with long lifetimes (more than 100 days). Therefore the effect of anabolic steroids is more noticeable in the muscle on contractile fibers with short lifetimes than in tendons on collagen with long lifetimes. This generates a disharmony between muscle and tendon, which can increase the probability of injury following training under the extensive use of anabolic steroids.

1.3 METABOLIC PROCESSES

Chemical reactions inside the cell are determined by physical and chemical laws of nature. The initial energy source is the sun; its electromagnetic light energy is consumed by plants via photosynthesis, and converted into chemical energy. At the first step of photosynthesis, a pigment molecule absorbs one photon and loses one electron, falling into a lower energy state. Electrons flow through an electron transport chain, leading to the ultimate reduction of NADPH. This creates a proton gradient, which is used by ATP synthase in the synthesis of ATP. In the second step of photosynthesis, which does not require light, the accessible NADPH and ATP are used for CO_2 fixation and the synthesis of carbohydrates. Higher-ranked organisms are not capable of capturing light energy, therefore they are dependent on energy produced by lower-ranked organisms. These organic molecules, proteins, carbohydrates, and lipids provide not only energy but also the keystones of cellular construction.

Animal and human cells obtain energy through the oxidation of organic molecules. Metabolism of nutriments, high energy carbohydrates, lipids, and proteins takes place in several steps. First, the food is digested by the digestive system resulting in free fatty acids from lipids, sugar molecules from carbohydrates, and amino acids from proteins. Further metabolism of these molecules with the usage of molecular oxygen provides energy resulting in final metabolites of carbon and hydrogen in the form of low energy carbon dioxide (CO_2) and hydrogen (H_2O).

Cellular oxidative metabolism is a coordinated step-by-step process unlike the oxidation outside an organism. Several enzymes are involved in this process, catalyzing reactions and participating in the best alternative path in a given condition. Biological oxidation does not mean the binding of oxygen to any molecules directly, but indicates an electron transport process, in which electrons flow through from one molecule to the next. The main energy sources are lipids and carbohydrates, which are oxidized to low energy molecules such as CO_2 and H_2O. This last step takes place in the mitochondria, where the energy will be stored in the form of ATP. ATP is readily used in cellular processes like bills in the shop. However carbohydrates and fat must be transferred to ATP to be able to use them. They are valuable, but we cannot use them readily to shop, but after converting them to into ATP can use them, so they are like gold and securities. Energy is provided via ATP hydrolysis, whereby chemical energy that has been stored in the high-energy phosphoanhydride bonds is released, for example, in muscles in the process of actomyosin binding or in transportation processes.

Carbohydrate metabolism may occur in the presence or absence of oxygen. This process, called glycolysis, takes place in the cytoplasm, where carbohydrates transform into pyruvate. From one sugar molecule, two molecules of pyruvate and two of ATP arise. The next step occurs in the mitochondria in the presence of oxygen. Upon the pyruvate entering the mitochondria, it converts to acetyl-CoA; this enters the Krebs-cycle, and at the end of the processes 36 ATP are produced altogether. In the case of a lack of oxygen—for example, in skeletal muscle sometimes—pyruvate does not enter the mitochondria but converts to lactic acid in the cytosol. Lactic acid is considered mainly a side product of anaerobic metabolism; however, it does also have important effects in the nervous system and immune system, and in vascularization.

Lipid metabolism requires more oxygen than that of carbohydrates. Production of acetyl-CoA is a step of the metabolism of free fatty acids generated by the digestion of lipids. The acetyl group of acetyl-CoA and its C——H and C——C bonds are oxidized and converted into carbon dioxide and H_2O, generating huge amounts of ATP stored in the phosphate bonds. From the generated carbon dioxide and hydrogen ions, the latter is picked up by NAD^+ coenzyme in the mitochondria. As electrons flow through the electron transport chain (see Fig. 1.3) in the inner membrane of the mitochondria, protons are pumped out to the intermembrane space, producing a gradient that serves as an energy source for the production of ATP. The total amount of ATP in a cell is approximately 10^9 molecules. Under normal circumstances, proteins are not sources of energy in a cell, but rather structural and functional units.

1.4 "R" EXTRA

Physical training increases the expression of several genes, posttranslational modification of histone proteins, which are responsible for keeping the DNA in a compact structure. Upon acetylation of histones by histone acetyl transferase enzymes, the chromosome opens up, making the transcription of DNA possible, and mRNA can be synthesized. An anaerobic environment, for example, strenuous physical training, activates hypoxia inducible factor 1 alpha (HIF–1 alpha), which is a transcriptional factor and translocates into the nucleus, inducing the expression of genes that are responsible for the defense of the cell under anaerobic conditions—for instance, enzymes that are involved in anaerobic metabolism, in which ATP can be generated in the absence of oxygen. HIF–1 alpha also induces genes involved in vascularization and mitochondrial biogenesis. Physical training at high altitudes thus increases the vascularization of musculature and improves the efficiency of oxygen combustion via the action of HIF–1 alpha.

Measurement of cellular adaptation can be performed by several methods. Upon tissue homogenization in different buffers that breaks up the tissue, cellular constituents can be separated by centrifugation at different speeds, therefore mitochondria, the nucleus, and other organelles can be examined separately. As a result, mRNA content and its protein products and protein activity are measurable in different cell compartments. The quantity of a protein of interest can be measured by antibodies, which are able to bond specifically to a particular protein. For instance, citrate synthase of the Krebs-cycle is an accepted protein reflecting the aerobic efficiency of physical training in skeletal muscle. It is also possible to measure the activity of an enzyme isolated from a tissue, which gives more important information compared to the quantity of a protein.

Mitochondrial biogenesis, an increase in the number of this organelle, is one of the most important events following physical training. It was discovered by a Hungarian-born scientist, John Holloszy, in 1968. This shed light on one possible path of cellular adaptation following physical training. The specific stimuli inducing mitochondrial biogenesis will be discussed in later chapters. Mitochondria develop a dynamic network; they are capable of fusion, and proliferation, which allows them to respond in a specific stimulus–response condition. Many processes are already known and others unknown that stimulates mitochondrial biogenesis. For instance, Ca^{++}, lack of oxygen, AMPK (adenosine monophosphate activated protein kinase), PGC-1 (peroxisome proliferator-activated receptor gamma coactivator

1-alpha, a transcription factor regulating metabolism), and free radicals are known stimuli that induce mitochondrial biogenesis.

A higher number of mitochondria not only supports more efficient ATP production but at the same energy demands can work at lower intensity, which influences the generation of free radicals. Lower intensity of mitochondrial work produces less, whereas higher intensity with higher amounts of oxygen consumed produces higher amounts of free radicals (Radak et al., 2013) (Fig. 1.8).

Side products of aerobic metabolism are free radicals, which are atoms or molecules possessing unpaired electrons. Reactive oxygen species are one of the most researched groups of free radicals; their production is in close proximity to metabolic processes. In a mitochondrial electron transport chain, where molecular oxygen is reduced to H_2O, every step produces free radicals. Oxygen molecules contain two unpaired electrons at 2p orbitals, making them susceptible to association with other molecules. This feature of oxygen makes it an excellent participant in chemical reactions. The generation of reactive oxygen species starts with the step of superoxide production, in which molecular oxygen (O_2) is converted to superoxide anion (O_2^-); it is approximately 1% of oxygen consumed by the mitochondria. Superoxide is highly reactive and is unable to pass through membranes; therefore if it "leaks out," it causes damage in close proximity to where it was generated—for

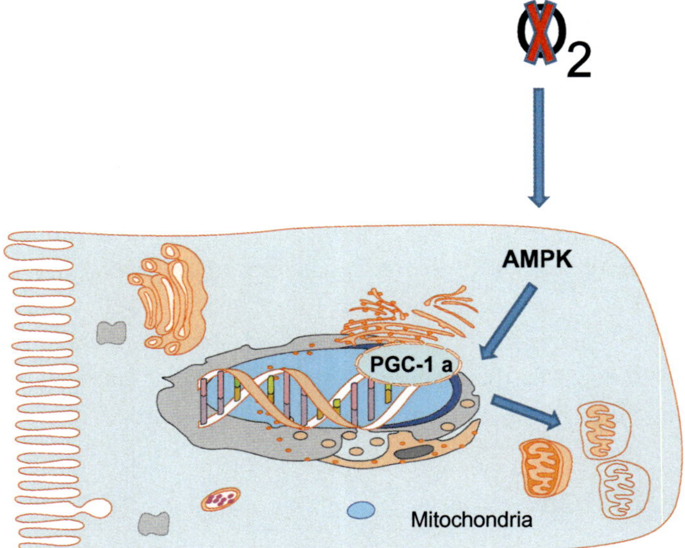

Fig. 1.8 Lack of oxygen and mitochondrial biogenesis. Low levels of oxygen or physical training induced mitochondrial biogenesis via the stimulation of PGC-1.

instance, reacting with lipid molecules of the membranes it neutralized in the form of H_2O. However, lipid peroxidation can be deleterious to membrane permeability, which in a serious form can lead to cell death.

Superoxide is converted to hydrogen peroxide (H_2O_2), which is able to pass through membranes. Hydrogen peroxide also serves as a signal molecule, for instance, it activates transcription factors and plays a role in stem cell differentiation. Moreover, it is able to bind to amino acids and modify their structure in the presence of iron. Hydrogen peroxide can convert into hydroxyl radical (OH^-), which is an extremely aggressive radical and able to harm DNA. The final step of free radical reduction is the conversion of hydroxyl radical to H_2O.

Free radicals leaking from the electron transport chain or radicals generated by enzymes are able to modify the structure of lipids, proteins, and DNA. This observation raised the free radical aging theory. The human body is armed against free radicals by antioxidant neutralizing systems. Moderate levels of oxidants are essential in cellular signaling, for instance, H_2O_2 helps in the opening of Ca-channels in muscle contraction, which is an important step of Ca-troponin association. Small amounts of hydrogen peroxide increase the force generation of muscle contraction, whereas high amounts of the oxidant play a role in muscle fatigue.

1.5 SUMMARY

The cell is the basic unit of all living organisms containing cell organelles and is limited by its cell membrane. Organelles are in dynamic relationships with each other. DNA encodes genetic information, which is localized in the nucleus and the mitochondria. Cells are able to respond to stimuli by modifying the activity of enzymes and protein expression. Cellular adaptation comprise of stimulus-dependent and independent processes. Cells produce energy from carbohydrates and lipids, and store this energy in the form of ATP, which can be used for cellular catabolic processes. This process takes place in the mitochondria. Carbohydrate metabolism can be oxygen-dependent (aerobic) and oxygen-independent (anaerobic), whereas lipid metabolism may occur only in the presence of ample oxygen.

TEST QUESTIONS

1. What are cellular membranes composed of, and what are the main cellular organelles?

2. Where is DNA located and what is its role?
3. What is the role of mitochondria in the cell?
4. What is the difference between aerobic and anaerobic metabolism?

BIBLIOGRAPHY

Ádám, V., Dux, L., Faragó, A., Fésüs, L., Machovich, R., Mandl, J., Sümegi, B., 2006. In: Ádám, V. (Ed.), Orvosi biokémia. Medicina Könyvkiadó Zrt., Budapest.

Fonyó, A., Hunyady, L., Kollai, M., Ligeti, E., Szűcs, G., 2003. Az orvosi élettan tankönyve. Medicina Könyvkiadó Zrt, Budapest.

Holloszy, J.O., 1967. Biochemical adaptations in muscle. Effects of exercise on mitochondrial oxygen uptake and respiratory enzyme activity in skeletal muscle. J. Biol. Chem. 242, 2278–2282.

Mooren, F., Volker, K. (Eds.), 2006. Molecular and Cellular Exercise Physiology. Human Kinetics, Champaign.

Radak, Z., Zhao, Z., Koltai, E., Ohno, H., Atalay, M., 2013. Oxygen consumption and usage during physical exercise: the balance between oxidative stress and ROS-dependent adaptive signaling. Antioxid. Redox Signal. 18, 1208–1246.

Sperelakis, N. (Ed.), 2001. Cell Physiology. Academic Press, San Diego.

CHAPTER 2

Skeletal Muscle, Function, and Muscle Fiber Types

This chapter discusses the characteristics of skeletal muscle, which determine sport performance and health. Skeletal muscle is capable of adaptation which is visible to the human eye—for example, muscle hypertrophy. The condition of the musculature determines our locomotion and our quality of life.

Skeletal muscle is specialized for locomotion and also has an important role in sugar and lipid metabolism, and affects the immune system. The two other types of muscle, cardiac muscle and smooth muscle, are not covered in this chapter. Each anatomically distinguishable muscle is surrounded by connective tissue called epimysium, and is built up by fascicles surrounded by perimysium. Muscle fascicles are built up by muscle fibers, which are limited by endomysium (Fig. 2.1).

Muscle fibers, which are also called myocytes, are specialized units since they have several nuclei, which are located under the sarcolemma. In the sarcomere, which is the basic functional unit of skeletal muscle, thin and thick myofibrils are responsible for locomotion and are arranged in a strict repeating order. Sarcomeres are divided by Z-lines (Fig. 2.2).

In the cytoplasm called sarcoplasm, there are mitochondria and other organelles such as sarcoplasmic reticulum, lipid droplets, and glycogen. The structure of a sarcomere is influenced by the differential content of slow-twitch and fast-twitch muscle fibers. Slow-twitch muscle fibers show thicker Z-lines and are less capable of hypertrophy compared to fast-twitch muscle fibers; however, they are more resistant to mechanical stress, and there are also some energetic differences. This has huge importance in sports—for instance, endurance sports require muscle fibers which are resistant to long-lasting mechanical stress, therefore slow-twitch fibers provide an advantage in these sport categories. Thin actin myofilaments, tropomyosin, and troponin complex attach to the Z-line; this area is called the I-band (Fig. 2.3).

In resting fibers, within the A-band, where thin and thick filaments overlap there is a paler area called the H-zone, which is the zone of thick

The Physiology of Physical Training
https://doi.org/10.1016/B978-0-12-815137-2.00002-4
 15

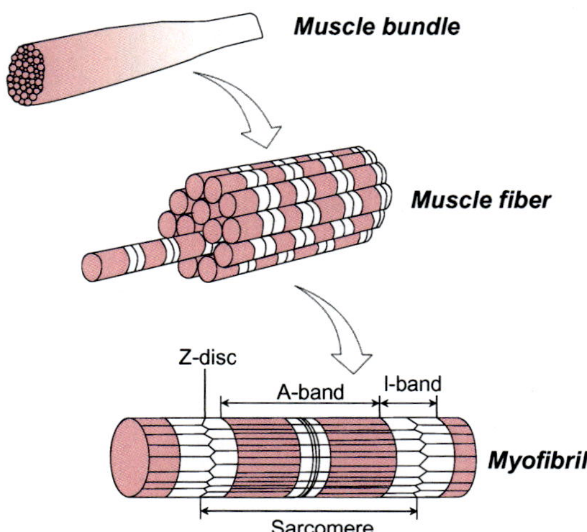

Fig. 2.1 Structure of skeletal muscle. The architecture of the skeletal muscle is perfectly structured anatomically.

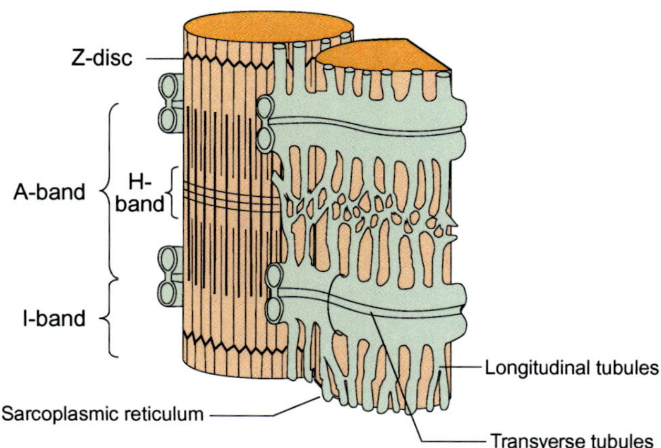

Fig. 2.2 Structure of muscle fibers. The sarcomere is the basic functional unit of skeletal muscle, which is responsible for muscle contraction.

filaments that is not superimposed by the thin filaments. In the middle of the sarcomere, myosin filaments are localized, and each myosin filament is surrounded by six thin actin filaments. During contractions, the width of I–bands and H–zones decreases, while during stretching it increases.

Myosin filaments were discovered by Albert Szent–Györgyi, while actin filaments were identified by his colleague, Brúnó F. Straub. Several other proteins are important for the stability of the sarcomeric structure. Titin is a giant protein, which is the biggest single highly elasticated protein found

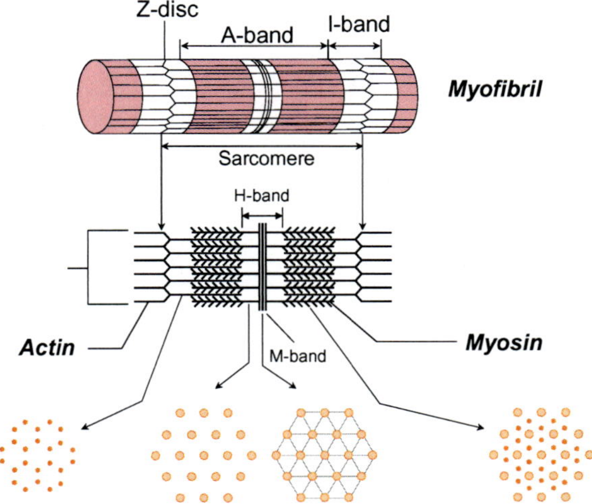

Fig. 2.3 Structure of a sarcomere.

in nature. It extends from the Z–line of the sarcomere to the M–band covering the half area of a sarcomere, and connected to several proteins such as calmodulin, alpha–actinin, and myosin-binding protein C. Absence of titin may lead to muscle atrophy. Another protein, nebulin, is also a giant protein; it attaches to troponin, tropomyosin, and calmodulin, and it guarantees the elasticity of a muscle (Fig. 2.4). Calmodulin is a small Ca binding protein, and maintains Ca homeostasis. Actin has many forms, such as alpha and beta actins, and these are important not only in muscle contraction but also as structural proteins. Dystrophins are structural proteins; the absence of this protein because of its mutated gene causes serious diseases such as

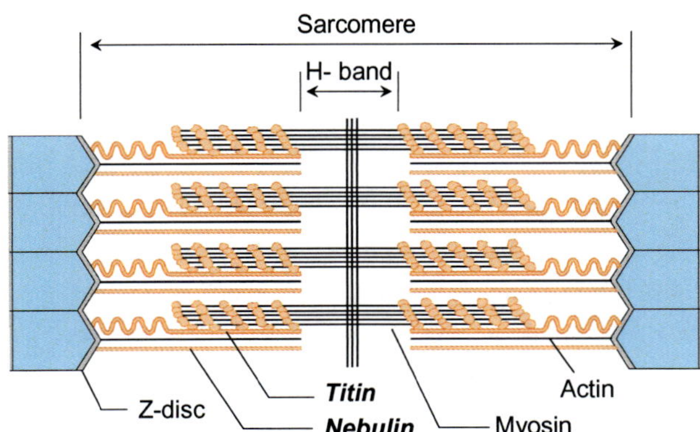

Fig. 2.4 Contractile and structural proteins in the sarcomere. Localization of titin and nebulin in the sarcomere.

Duchenne or Becker syndromes with muscle atrophy. Dystrophin binds neural nitrogen oxide synthase (nNOS), which regulates muscle functions including muscle hypertrophy or reparation after muscle injury.

Based on the arrangement of muscle fibers, muscles show different architectures. The fascicles of longitudinally arranged parallel muscles run parallel to the axis of force generation, and they have a ribbon-shaped longitudinal architecture; they connect to their tendons in a wide area. The muscle-tendon junction is the weakest area in the musculature, and the tension that a muscle can create between its tendons may result in injuries. The middle part of the longitudinally arranged fusiform muscles is called the muscle belly, and shortens and thickens upon contraction. Most muscles in the human body belong to this group. The sarcomere is the basic functional unit of skeletal muscle, and can shorten by 30%; therefore considering a 10 cm contraction, it shortens the length of a muscle by 3 cm. The tension that a longitudinally arranged muscle can create between its tendons is approximately $3.6 \, kg/cm^2$.

The fibers in triangle-shaped muscles (e.g., pectoralis major) converge at one end and spread over a broad area at the other end. They have a weaker pull on the attachment side compared to other parallel fibers, and thus generate less force.

Unlike in parallel muscles, pennate fibers (unipennate, bipennate, multipennate) are at an angle to the force-generating axes. This angle reduces the effective force of any individual fiber; the fibers cannot shorten as much as in longitudinally arranged muscles. However, because of this angle, more fibers can be packed into the same muscle volume, enabling considerable force to be generated.

The fibers in unipennate muscles are all oriented at the same angle relative to the axis of force generation (e.g., the extensor digitorum muscles in the forearm). In the more common bipennate muscles (e.g., the rectus femoris) there are fascicles of both sides of the central tendon facing in opposite diagonal directions. Multipennate muscles (e.g., deltoid muscle) are constructed of several overlapping muscle fibers running at different angles (Fig. 2.5). Circular muscles (e.g., the orbicularis oris encircling the mouth) appear circular in shape and are normally sphincter muscles which surround an opening; during contraction this muscle narrows the opening.

Although muscles are the focus of contraction, tendons have special importance in locomotion since they connect muscle to bones, and transmit forces resulting in locomotion. The optimal load based on physiological and structural differences of skeletal muscle, connective tissue, and tendons will be discussed in later chapters.

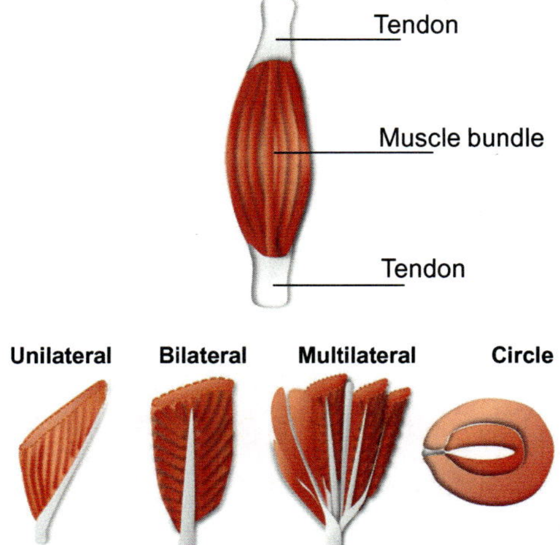

Fig. 2.5 Arrangements of muscle fibers. Based on the arrangement of muscle fibers, muscles show different architectures resulting in different shortening and force generation.

2.1 MUSCLE CONTRACTION

The axons of the neuromuscular junctions originate from the motor neurons located in the ventral horn of the spinal cord. The motor neuron and all the muscle fibers to which it connects is a motor unit. The neuromuscular synapse comprises of postsynaptic and presynaptic membranes. The presynaptic membrane is the axon terminal containing acetylcholine vesicles. Upon stimulation, the stored acetylcholine pool is released and binds its receptor on the postsynaptic membrane of the sarcolemma, causing ion channels to open, and allows sodium ions to flow across the membrane into the muscle cell. This generates an action potential which travels to the myofibril through the transverse tubule system, and the sarcoplasmic reticulum; consequently Ca-ions release, which results in muscle contraction. Ca-ions bind to troponin C, resulting in conformational changes, which allow myosin to bind to actin, producing muscle contraction. In a resting state, troponin-I of the troponin complex covers the actin-myosin contact area. Actin-myosin binding requires ATP, which binds to the myosin-heavy chains (head) (Fig. 2.6).

In the resting state of a muscle, myosin has ADP and Pi (inorganic phosphate) bound to its nucleotide binding pocket. In the first step of the actin-myosin binding process, inorganic phosphate is released, followed by a power stroke and the release of ADP. This will pull the Z-lines toward each

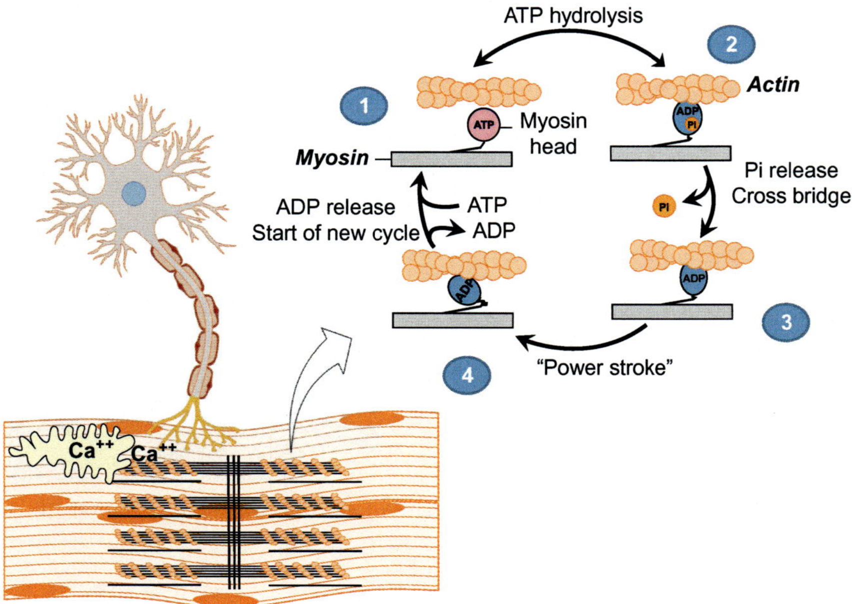

Fig. 2.6 Molecular mechanism of muscle contraction. Action potential travels through the transverse tubule system and the sarcoplasmic reticulum, resulting in Ca-ions release. Ca-ions bind to troponin C resulting in conformational changes, which allow myosin to bind to actin, producing muscle contraction. During each actin-myosin binding, protein filaments slide on each other to produce a contraction, which requires ATP.

other, thereby shortening the sarcomere, approximately 10–12 nm/stroke. In the next step, ATP binds to the myosin, which weakens the attachment between actin–myosin filaments, in turn allowing the release of actin and the break of the cross bridge. If the level of ATP is low, it may result in a contracture.

Myosin heads, which are involved in the formation of the cross bridges, have different characteristics in slow-twitch and fast-twitch muscle fibers. In fast-twitch muscle fibers, formation of the cross bridges between actin and myosin filaments occurs faster because of the higher ATPase activity of myosin, and also the binding capacity of their binding sites are significantly high. Following contraction, Ca-ions are transported back to the sarcoplasmic reticulum by active transport, and troponin C returns to its resting state, so the muscles are able to relax.

Fast strokes demand a great deal of ATP molecules, which challenges the metabolic capacity of the body. To create significant force, more cross bridges need to be formed, whereas fast movements require cyclic changes of cross bridges and high amounts of ATP. Thus, fast-twitch muscle fibers show high ATPase activity, which was described for the first time by a Hungarian scientist, Mihály Bárány. Slow muscle movements require fewer cross bridges to be formed in a given time frame, reducing the ATP demand. This explains why huge amounts of ATP are required by maximal velocity movements, and why lower amounts of ATP are necessary for maximal force movements and much less for endurance movements. This topic will be further discussed in the next chapters.

2.1.1 Types of Contractions

There are three types of muscle contraction: concentric, isometric, and eccentric. Labeling eccentric contraction as "contraction" may be a little misleading, since the length of the sarcomere increases during this type of contraction. Thus in this context contraction does not necessary imply shortening (Table 2.1).

- In a concentric contraction, the force generated by the muscle is less than the muscle's maximum, and the muscle begins to shorten. This type of contraction is widely known as muscle contraction. It requires more energy compared to the other two types, but this contraction generates the least force.
- An isometric contraction generates force without changing the length of the muscle, and no mechanical work is done since the muscle does not shorten. However, this type of contraction requires high amounts of energy because of the force generated by the muscle. This force is equal to the external load, thus the length of the muscle does not change.
- In an eccentric contraction, the external force on the muscle is greater than the force that the muscle can generate, thus the muscle is forced to lengthen due to the high external load. The maximal force generated by the muscle is the highest; however, the energy consumption is the lowest.

Comparison of maximal force generation in concentric, isometric, and eccentric contractions show the following ranking: eccentric > isometric > concentric. This ranking can be explained by muscle-tendon characteristics.

Table 2.1 Types of contraction

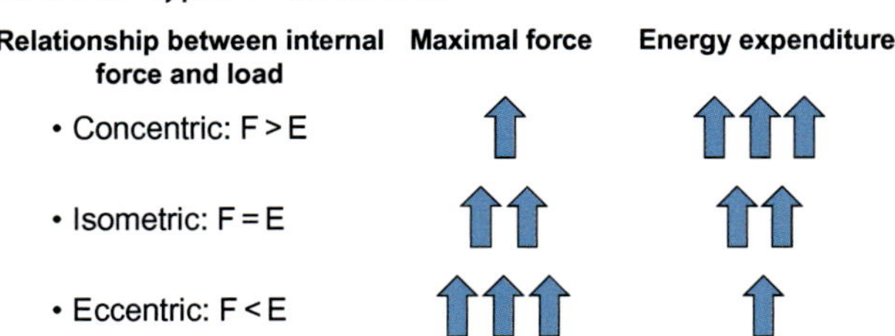

Relationship between internal force and load	Maximal force	Energy expenditure
• Concentric: F > E	↑	↑↑↑
• Isometric: F = E	↑↑	↑↑
• Eccentric: F < E	↑↑↑	↑

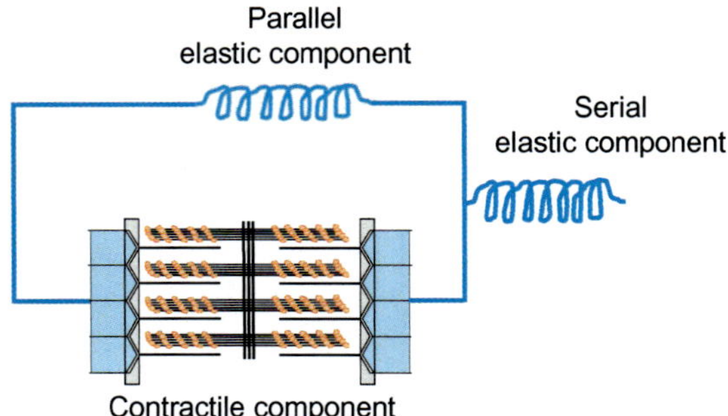

Fig. 2.7 Hill's three-element model of muscle contraction. Force is generated by the contractile element, while the elastic elements store the energy.

Archibald V. Hill is the only scientist who received a Noble-prize for his work done on sport-related research: the mechanical work in muscles. Hill's three-element model is a representation of the muscle mechanical response. The model is constituted by a contractile element, a series element, and a parallel element (Fig. 2.7). The contractile element comes from the force generated by the actin and myosin myofibrils cross bridges. The series element represents the tendons, which are extensile but as much as muscles. The parallel element represents the passive force of the connective tissues such as fascia, membranes (epimysium, perimysium, endomysium), titin, and nebulin. In concentric muscle the force is generated collectively by the contractile element and at less extent the parallel element, while isometric muscle contraction involves the full contribution of the parallel element.

In an eccentric contraction, the external force on the muscle is greater than the force that the muscle can generate, thus the series elastic elements are also forced to lengthen due to the high external load. This mechanical energy can be reclaimed during contraction, which provides a higher force. The model in Fig. 2.7 is a simplified explanation of why the different muscle contraction types produce different amounts of maximal force (Fig. 2.8).

2.2 TYPES OF MUSCLE FIBERS

Slow-twitch and fast-twitch muscle fibers have already been addressed briefly; here we discuss them in detail. Muscle is comprised of different muscle fibers, which differ in appearance and other characteristics. For instance, comparing muscle isolated from a wild and a domestic rabbit, the wild one is more reddish in color. Also when comparing chicken breast to thigh, the latter is more reddish than the breast.

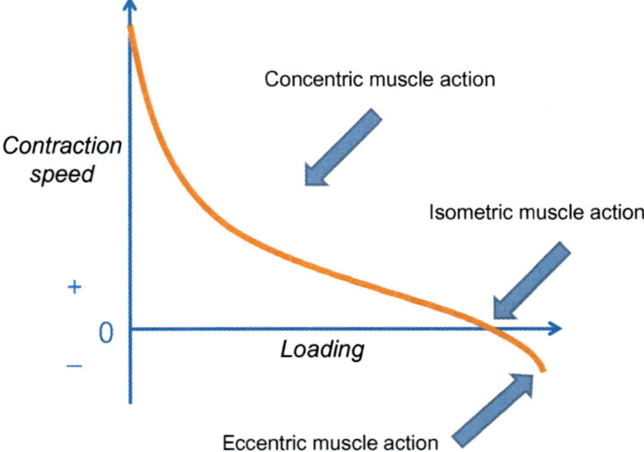

Fig. 2.8 Force-velocity relationship. A.V. Hill's force-velocity curve shows that the speed at which a muscle changes length also affects the force it can generate. The shortening velocity increases as the force declines, and so increasing force results in the decline of shortening velocity. If the force further increases the muscle is not able to shorten further; it contracts isometrically. If the external force on the muscle is greater than the force that the muscle can generate, the velocity turns to negative, in the case of eccentric muscle contraction.

Thus, muscle that is exposed to constant activity (such as the muscle of a wild rabbit, and the thighs of the chicken) is reddish, and composed of slow-twitch muscle fibers, whereas muscles which are not exposed to constant exertion (muscle of a domesticated rabbit, and chicken breast) are lighter in color and are composed of fast-twitch muscle fibers. The occurrence of muscle fibers depends on exertion, innervation, and the type of innervation. Inside a muscle there may appear different type of muscle fibers—for instance, closer to the bones, muscles are more reddish than close to the surface. Generally speaking, extensors contain more fast-twitch muscle fibers than flexors do. There are muscles in the human body that are comprised mainly of either slow-twitch or fast-twitch muscle fibers. Muscle fibers are innervated by alpha motor neurons. The motor neuron and all the muscle fibers to which it connects is a motor unit. The number of muscle fibers innervated by one motor neuron can differ; for instance, in extraocular muscles, 10 muscle fibers are innervated by one motor neuron, while thigh muscles can have 1000 fibers in each unit. The axons of the motor neurons of the spinal cord innervate the peripheral muscles, and they can have a length of more than 1 m (Fig. 2.9).

Motor units differ according to size and activation threshold. Large motor units have higher activation thresholds and contain paler fast-twitch

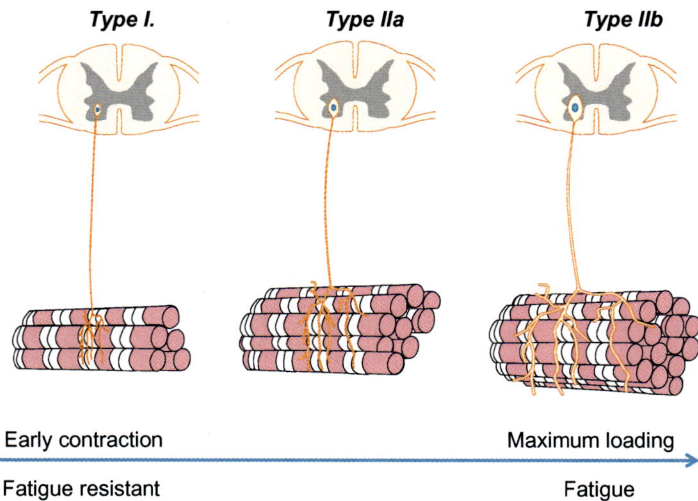

Fig. 2.9 Motor units. There are three different motor units in the human body. The type I motor unit is highly fatigue resistant, has a lower activation threshold, contains fewer muscle fibers, and has low force generation during contraction. The type II motor unit is also resistant to fatigue, has a higher activation threshold, and the force produced is higher compared to type I. The type IIb motor unit is fatigable, has a high activation threshold, innervates the most muscle fibers, and generates the greatest force during contraction.

muscle fibers, while small motor units have lower activation thresholds and contain reddish slow-twitch muscle fibers. Differences in physiological, biochemical, histochemical, and genetic characteristics of motor units also provide a useful base to distinguish between them (Table 2.2).

Based on physiological characteristics, human muscle fibers are fast fatigable, fast fatigue resistant, fast intermediate, or slow fibers; based on biochemical properties they are fast glycolytic type IIb fibers, fast oxidative-glycolytic type IIa fibers, or slow oxidative type I fibers. Another classification gives another type of fibers, IIi fibers, with characteristics between type IIa and IIb. The red slow-twitch fibers contain high amounts of iron, which come from the higher number of mitochondria and myoglobin content. The red fibers have higher oxidative capacity; they are able to consume high amounts of oxygen and reduce it in the mitochondria. Oxygen is always bound to iron-containing molecules; this high iron content also contributes to its red color. Table 2.2 shows the differences between the types of muscle fibers. The differences in activation threshold determine the activation order of the different fiber types in contraction. Slow-twitch muscle fibers with their excellent oxygen consumption rate and high mitochondria content

Table 2.2 Characteristics of muscle fibers

Characteristics	Fast-twitch fibers	Slow-twitch fibers
Time to maximal contraction (ms)	50–80	100–200
Frequency to reach tetanic contraction (Hz/s)	60	16
Myoglobin and mitochondria density	Low	High
Dominant pathway of ATP synthesis	Anaerobic	Aerobic
Glycogencontent	High	Low
Myosin-ATPase activity	High	Low
Capillarization	Low	High
Fatigue resistance	Poor	High
Size of motor neuron	Big	Small
Threshold	High	Low
Force-generating capacity	High	Low

and oxidative enzyme activity are the most efficient fibers. They are able to generate force at the point of contraction because of their low activation threshold. Most fibers of antigravity muscles are slow-twitch fibers, and these fibers are involved during walking and low intensity movements. One of the main laws in nature is profitability, which in this case means the engagement of the more profitable muscle fibers first. Fast-twitch muscle fibers with their higher activation thresholds and huge force generation are usable during flight and survival; however, these fibers consume a lot of energy and produce a lot of lactic acid (discussed later). They can only be activated by high intensity stimuli because of the higher threshold. To use an analogy, slow-twitch fibers are like economic city cars, while fast-twitch muscle fibers are like high-powered racing cars.

2.3 TENDONS AND CONNECTIVE TISSUE

With the exception of mimetic facial muscles, muscles are connected to bones by tendons, which comprise of collagen fibers. Collagen has slow turnover (balance between protein synthesis and protein degradation), and a high half-life (over 200 days). Thus the adaptation of tendons is not comparable to muscle plasticity; however, their flexibility can be improved following physical training and can become more resistant to loading. Collagen is 85% of the dry weight of a tendon. A tendon's oxygen and blood supply is moderate, healing following and injury is slow, and aging of tendons is faster because of the accumulation of the free radical related injuries. As already mentioned, the muscle-tendon junction is the most vulnerable part of the musculature, because it involves the connection of two totally different structures.

2.4 SKELETAL MUSCLES AND AGING

Aging affects the muscular system after the age of 50, and by the age of 80 the system may have lost 50% of its motor units. The loss of motor neurons and muscle fibers occur simultaneously, thus the cause has the same root. Deterioration of fast-twitch fibers occurs faster than that of slow-twitch fibers. A loss in muscle weight not only affects mobility but also has an adverse effect on blood sugar regulation since uptake of sugars by the muscle from the blood is insulin dependent, so less sugar can be taken up by the decreased capacity of the muscle by a given amount of insulin. Insulin resistance is reflected by a decreased uptake of sugar by the muscles from the blood.

The result or the cause of the loss of muscle weight in the elderly is impaired physical activity. However, physical training would be necessary to decelerate decalcification of bones, which is also affected by the loss of muscle weight. Frequent physical activity may decelerate deterioration of the musculature; however, it cannot prevent it completely.

2.5 "R" EXTRA

2.5.1 Fiber Type Assessment

The Bergstrom muscle biopsy technique has been an essential tool for the direct assessment of human skeletal muscle. It is an invasive procedure in which a 10–50 mg piece of muscle tissue is removed from the muscle under local anesthesia. The procedure can be performed several times with the use of a smaller biopsy needle, which gives an advantage to following metabolic processes (e.g., glycogen metabolism during training).

Microscopic assessment of the paraffin-embedded samples can be performed via different staining procedures (e.g., ATPase, NADH, tetrazolium staining). A higher rate of slow-twitch muscle fibers (over 60%) indicates an advantage in endurance sports, whereas a higher rate of fast-twitch muscle fibers is more beneficial in power sports. MRI (magnetic resonance imaging) can also be used for the assessment of muscle types.

Although the types of muscle fibers are determined by genetic factors, stimuli have some effect on them. For instance, changing the type of innervation by surgical procedures has a huge effect on the metabolic properties of a given muscle fiber. The same results have been observed by electrical muscle stimulation, in which stimuli specific to fast-twitch muscle fiber stimulation were applied to slow-twitch muscle fibers, and vice versa. Thus the type of physical training has a tremendous effect on the metabolic processes of muscle fibers, which can be exerted in the case of intermediate fibers.

Interestingly, metabolism of AMP in fast-twitch muscle fibers produces ammonia by AMP-deaminase activity, and the presence of ammonia in the blood and the central nervous system results in fatigue. In contrast, AMP metabolism takes the course of adenosine rather than inosine pathway by the enzyme 5′ nucleotidase, resulting in an absence of ammonia in slow-twitch muscles.

Mechanical immobilization following injury also has a major effect on the type of muscle fibers, especially on slow-twitch muscle fibers. These fibers are activated most of the time because of their low activation threshold, whereas fast-twitch fibers are generally inactive because of their high activation threshold. In contrast, loss of fast-twitch muscle fibers occurs in the elderly, whereas slow-twitch muscle fibers are not greatly affected. Characteristics of fast-twitch muscle fibers change over time, and their biochemical and physiological properties become similar to those of slow-twitch muscle fibers, which is detectable by the loss of power.

During muscle contraction, electrical activity of muscles can be measured by electromyography (EMG), where electrodes are placed on the surface or electrode needles are placed in the muscle. This method provides useful information about the activation of muscles.

2.5.2 Molecular Markers in Skeletal Muscle

Assessment of muscle fibers led to the discovery of proteins, which play a role in the development of different muscle fibers in animals used in laboratory experiments. The type of contraction regarding duration, tension, and hormonal changes is also able to affect the metabolism and maybe the composition of a muscle. Upon frequent contraction, the ATP level is decreased and consequentially the AMP level is elevated, which activates adenosine monophosphate protein kinase (AMPK), leading to stimulation of PGC-1 alpha (peroxisome proliferator-activated receptor gamma coactivator 1-alpha). PGC-1 alpha stimulation results in mitochondrial biogenesis. AMP is an important regulator of cellular adaptation since AMPK is involved in carbohydrate uptake and lipid oxidation, and also through the inhibition of protein expression initiated by IGF-1, or the activation or inhibition of proteins necessary for hypertrophy. PGC-1 activation is also regulated by Ca^{++}, which activates the calmodulin-calcineurin system.

Under the basal lamina in the skeletal muscle there are satellite cells (stem cells), which are able to divide and differentiate upon injury or following contraction. AMPK activation by, for instance, AICAR, a synthetic compound mimicking AMP, stimulates cellular processes that promote oxidative metabolism, lipid and carbohydrate metabolism, and inhibit stem cell proliferation and differentiation.

One of the main functions of skeletal muscle is the regulation of blood sugar level, which is regulated through insulin and mitogen-activated protein kinases (MAPK). Keeping the blood sugar level in an optimal zone is an important element of homeostasis. A modern lifestyle with a lack of physical activity may result in the development of type II diabetes, which is preceded by insulin resistance. Insulin is produced by pancreatic beta cells, which travel through the bloodstream and bind to their receptor and stimulate GLUT4 and its transport to the membrane in muscle. GLUT4 in the membrane is capable of uptaking the glucose molecules, resulting in a drop in blood sugar level. The MAPK system increases the level of GLUT4, thereby improving the muscle's uptake capacity of glucose and consequentially preventing the development of diabetes. Physical inactivity or aging result in a decrease in muscle weight and consequential decrease a muscle's uptake capacity of glucose, which in turn increases the incidence of type II diabetes (Fig. 2.10).

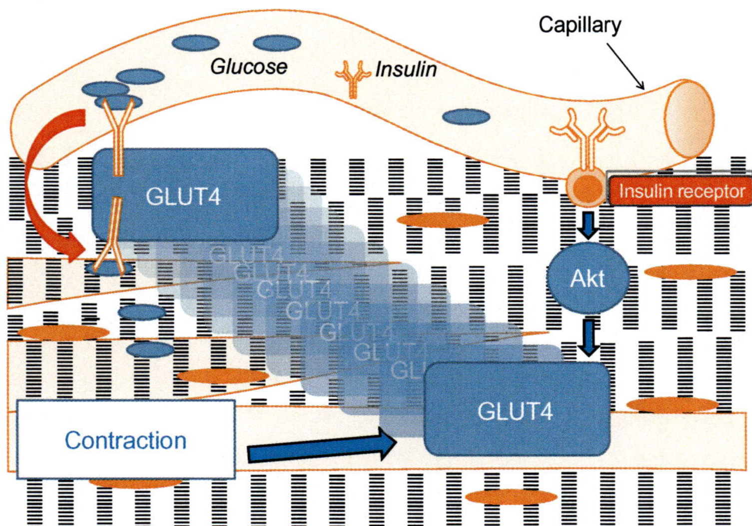

Fig. 2.10 Insulin-dependent uptake of glucose in muscle. Skeletal muscle is the largest organ uptaking sugar from the blood. Insulin upon binding to its receptors activates Akt, which results in the translocation of GLUT4 glucose transporter to the membrane. This makes the muscle capable of the uptake of glucose. Muscle contraction itself is sufficient to stimulate that translocation, and the muscle is thus able to uptake glucose from the blood during physical exercise.

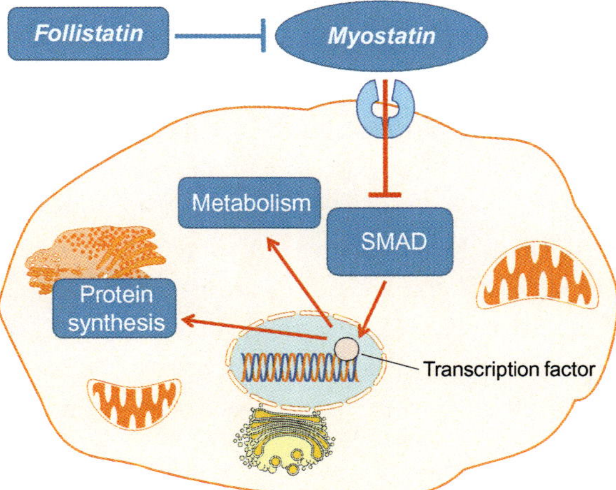

Fig. 2.11 Inhibition of myostatin. Follistatin inhibits myostatin, which is a negative regulator of muscle growth. Follistatin thereby induces protein expression in skeletal muscle, which is reflected by cross-sectional growth of the muscle.

Myostatin is a member of the transforming growth factor-beta protein family, and it is a negative regulator of muscle growth, thus its decreased activity results in muscle growth and improved reparation following injury. Knockout of myostatin gene in experimental animals resulted in increased mass of muscle and bone, and metabolic states were also altered since the mass of the fat tissue also decreased (Fig. 2.11).

Individual differences in myostatin gene activity may be possible, which would partly explain sport performance differences. Genetic analysis of racehorses showed differences in myostatin gene polymorphism (gene variations) (Hill et al., 2010). Follistatin is an endogenous inhibitor of myostatin; it has a role in embryonal development and differentiation, which is stimulated by nitrogen oxide (NO) and cyclic guanosine-monophosphate (cGMP). Follistatin-288 is a synthetic inhibitor of myostatin, and has been tested as a therapeutic of skeletal muscle atrophy; however, long-term effects and side effects have not yet been investigated. Myostatin inhibitors are considered to be doping drugs in sports.

2.6 SUMMARY

The architecture of muscle fascicles determines the force that a muscle can generate. Muscles are connected to bones by tendons, and the muscle-tendon junction is the weakest area of the musculature, making it vulnerable to injuries. The sarcomere is the basic functional unit of skeletal muscle, limited by two Z-lines, which is the place of actin myofibrils attachment. During muscle contraction, Ca-ions are released from the sarcoplasmic reticulum and bind to troponin C, resulting in its conformation change. This leads to attachment of myosin and actin filaments, consumption of ATP, and consequential muscle shortening. Muscle contraction can be categorized as concentric, isometric, or eccentric, and the type of contraction determines the power that the muscle can generate. A single motor neuron and all the muscle fibers to which it connects is a motor unit; these units are categorized as type I red fibers, type IIa intermediate, and type IIb fast fibers based on their physiological properties. Muscle fibers can be categorized differently depending on their structural, metabolic, and stimulation properties.

TEST QUESTIONS

1. What are the main components of a myocyte?
2. What categories of muscle architecture can be established based on the arrangement of muscle fibers? What is the role of the arrangement of muscle fibers regarding the extent of muscle contraction?
3. What type of muscle contractions can be defined and what are their characteristics?
4. What is the mechanism of muscle contraction?
5. What is a motor unit?
6. What are the characteristics of fast, slow, and intermediate muscle fibers?

BIBLIOGRAPHY

Aagaard, P., Suetta, C., Caserotti, P., Magnusson, S.P., Kjaer, M., 2010. Role of the nervous system in sarcopenia and muscle atrophy with aging: strength training as a countermeasure. Scand. J. Med. Sci. Sports 20, 49–64.

Bosco, C., Tihanyi, J., Komi, P.V., Fekete, G., Apor, P., 1982. Store and recoil of elastic energy in slow and fast types of human skeletal muscles. Acta Physiol. Scand. 116, 343–349.

Glass, D.J., 2010. Signaling pathways perturbing muscle mass. Curr. Opin. Clin. Nutr. Metab. Care 13, 225–229.

Hill, E.W., Gu, J., Eivers, S.S., Fonseca, R.G., McGivney, B.A., Govindarajan, P., Orr, N., Katz, L.M., MacHugh, D.E., 2010. A sequence polymorphism in MSTN predicts sprinting ability and racing stamina in thoroughbred horses. PLoS One 5, e8645.

Hortobagyi, T., Houmard, J.A., Stevenson, J.R., Fraser, D.D., Johns, R.A., Israel, R.G., 1993. The effects of detraining on power athletes. Med. Sci. Sports Exerc. 25, 929–935.

Kimball, S.R., Jefferson, L.S., 2010. Control of translation initiation through integration of signals generated by hormones, nutrients, and exercise. J. Biol. Chem. 285, 29027–29032.

Linklater, J.M., Hamilton, B., Carmichael, J., Orchard, J., Wood, D.G., 2010. Hamstring injuries: anatomy, imaging, and intervention. Semin. Musculoskelet. Radiol. 14, 131–161.

Malisoux, L., Francaux, M., Theisen, D., 2007. What do single-fiber studies tell us about exercise training? Med. Sci. Sports Exerc. 39, 1051–1060.

Moyes, C.D., LeMoine, C.M., 2005. Control of muscle bioenergetic gene expression: implications for allometric scaling relationships of glycolytic and oxidative enzymes. J. Exp. Biol. 208, 1601–1610.

Olesen, J., Kiilerich, K., Pilegaard, H., 2010. PGC-1alpha-mediated adaptations in skeletal muscle. Pflugers Arch.: Eur. J. Physiol. 460, 153–162.

Radak, Z., Chung, H.Y., Goto, S., 2008. Systemic adaptation to oxidative challenge induced by regular exercise. Free Radic. Biol. Med. 44, 153–159.

Sargeant, A.J., 2007. Structural and functional determinants of human muscle power. Exp. Physiol. 92, 323–331.

Schiaffino, S., Sandri, M., Murgia, M., 2007. Activity-dependent signaling pathways controlling muscle diversity and plasticity. Physiology 22, 269–278.

Sun, Z., Liu, L., Liu, N., Liu, Y., 2008. Muscular response and adaptation to diabetes mellitus. Front. Biosci. J. Virtual Libr. 13, 4765–4794.

Tee, J.C., Bosch, A.N., Lambert, M.I., 2007. Metabolic consequences of exercise-induced muscle damage. Sports Med. 37, 827–836.

CHAPTER 3

Adaptation, Phenotypic Adaptation, Fatigue, and Overtraining

3.1 HOMEOSTASIS AND ADAPTATION

Adaptation is the greatest characteristic necessary to survival of living organisms. Adaptation in a wider sense includes structural, morphological, and genetic changes upon environmental impact (sportgenomics), metabolic changes of organs (heart rate in untrained and trained athletes), or alteration of tactics in a team or as an individual.

Lamarck was the first to describe the correlation between species and environment in the 18th century; later Darwin shed light on the necessity of constant alterations in response to environmental changes, and the role of genetic and physiological fitness in ontogenesis.

In fact, sport science is all about adaptation at different levels (selection), scientifically planned modifications (training), and the selection of the best alternatives (competition). It is also true to all life, as it is the sum of successful and unsuccessful adaptations to environmental and social changes. In sport science, adaptation is didactically divided into several parts, which will be addressed in this chapter. To understand adaptation, we first need to discuss homeostasis.

Homeostasis implies a relatively stable equilibrium, which indicates constant changes in response to internal (e.g., pulse increase upon training) and external (e.g., environmental temperature) factors. Homeostasis variables are regulated in an optimal zone, and beyond this zone the chance for survival is low. For instance, body temperature may withstand 15–42°C; however, optimally it is 36.6–36.8°C. It can tolerate 34–39°C, but adaptive reactions are to restore the optimal temperature by shivering to generate heat or sweating to decrease body temperature. The optimal zone (e.g., blood sugar 3.6–5.8 mmol/L, blood pH 7.35–7.45) is very important to maintain physiological and biochemical processes of the body; for instance, low blood sugar, or hypoglycemia, induces glucagon, while high blood sugar stimulates

   33

insulin response. Homeostasis is an essential part of survival. Physical training can induce significant differences in individuals through adaptation processes; for instance, untrained athletes during an endurance running test stop activity when lactic acid in their blood reaches 8–10 mmol/L, whereas trained athletes keep running over 20 mmol/L. Thus, homeostasis stands for internal dynamic equilibrium with optimal levels, which may vary between species. Beyond the optimal zone, psychological, physiological, morphological, and other adaptational processes determine further possible individual differences. The next paragraphs will discuss adaptation in detail.

Certain skills, abilities, and anthropometric features are mainly determined by genetic factors; however, they are also influenced by environmental factors, resulting in phenotypic adaptation. For instance, the ratio of muscle fiber types is determined by genetic factors, based on studies of twins. The ratio of muscle fibers has an importance in individual sport selection. Genetic determination is influenced by geographic locations (e.g., adaptation to high altitudes), and this influence manifests clearly, for instance, in the outstanding performance of East African athletes in middle- and long-distance running. In contrast, West African and African Americans with genetic roots from West Africa are successful in power sports, and not that successful in endurance sports, indicating the role of genetic factors in addition to cultural factors. In the chapter about sportgenetics, the role of genes in sport performance will be discussed in detail since it has a huge impact on professional sports.

The human body is able to adapt to environmental stimuli and react specifically. General adaptation was first described by Janos Selye, a Hungarian scientist. He investigated the reaction of the thymus and the adrenal gland to different compounds injected into animals, and observed the exact same reaction regardless of the characteristics of the injected material. The size of the thymus is decreased, whereas the size of the adrenal gland is increased. He also described the behavioral changes in addition to the morphological changes. The concept of stress has recently become a very important theory, and it has a special importance in sports. Selye's general adaptation syndrome (Fig. 3.1) shows that continuous exposure to stress factors leads to exhaustion, which may have fatal consequences. Stress reaction is not a specific reaction, but rather a general reaction. In contrast, adaptation following physical training is a specific reaction.

Regarding physical training, we need to distinguish adaptation following a single training and regular training. A single, acute exercise bout induces reactions similar to those described by general adaptation theory. In contrast,

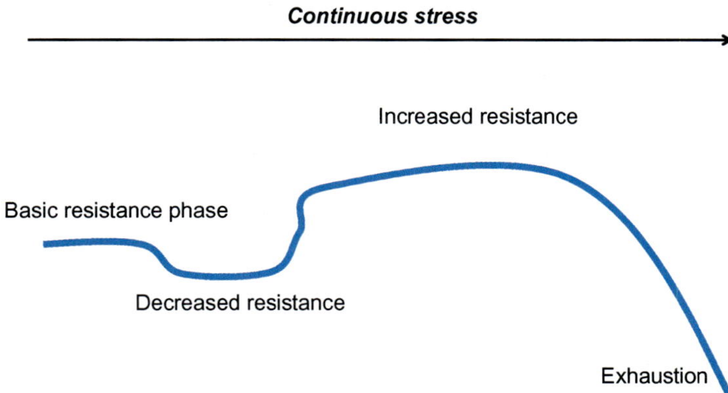

Fig. 3.1 Selye's general adaptation theory. Selye's general adaptation syndrome describes general reactions independent of the type of stressors. In the first stage, the stressor upsets homeostasis or cellular balance; in the next stage, the body fights back by adjusting to the stress. In the final stage, continuous exposure to the stressor leads to exhaustion, which may have fatal consequences.

regular training that includes resting intervals trains the body to adjust to the stress via metabolic, morphological, or psychological changes. Following a single acute loading, the adaptation may focus only on sweating, lactic acid elimination, and mobilization of fat sources. In contrast, adaptation following regular training causes phenotypical alterations to handle recurrent stress. For instance, a single, acute bout of exercise decreases the glycogen content of the muscle, whereas with regular training, muscles accumulate glycogen.

The intensity and duration of the training, and the resting periods, have a significant effect on the adaptation process, resulting in a specific adaptation response, in contrast to the general adaptation syndrome described by Selye.

The stressors or stimuli in physical training need to be frequent with optimal amplitude and duration. Suboptimal intensity does not induce adaptation (e.g., walking to get the remote control). If intensity exceeds the appropriate level, it may induce changes that result in longer regeneration time, which may break the continuity (in Greek history, the first marathon runner who delivered the outcome of a battle is claimed to have died following that single acute physical loading).

Adaptation following physical training is described nicely by Matveyev's periodization model. A modified, more didactic model can be seen in Figs. 3.2 and 3.3.

As mentioned earlier, the intensity of a stimulus is a key characteristic in the adaptation process. During a single acute episode of training, adaptation

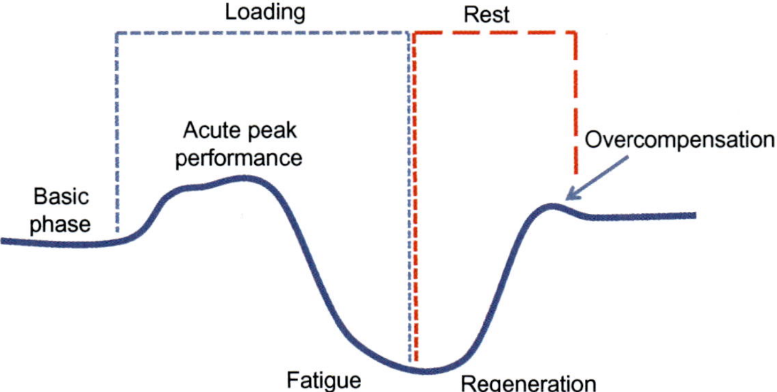

Fig. 3.2 Adaptation following a single bout of exercise. During training, the level of performance increases compared to resting state, and upon fatigue it starts to decline. In the resting state, body regenerates. Overcompensation may occur following frequent training, or sometimes following a single bout of exercise. For instance, the body can tolerate a higher lactic acid concentration and perform well following anaerobic training.

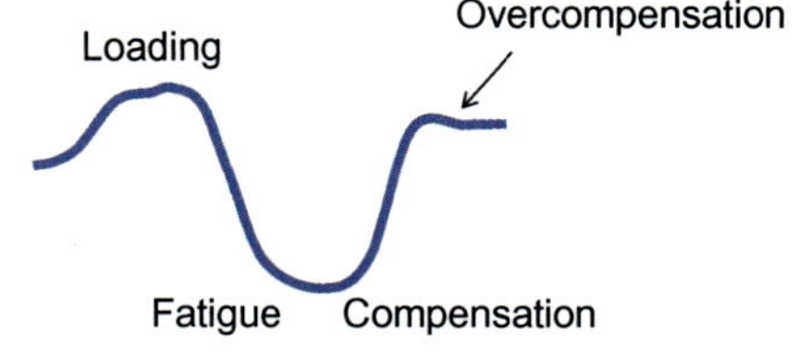

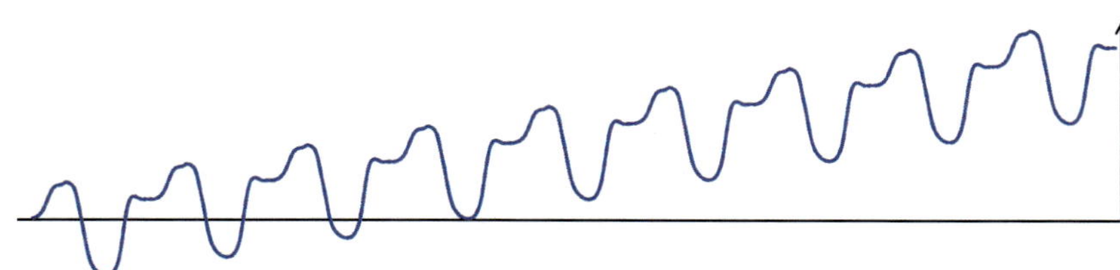

Fig. 3.3 Adaptation following regular training. Optimal training intensity and duration with optimal resting periods improve performance; this occurs sooner in the case of a trained athlete compared to an untrained one.

is limited since adaptation begins during the resting period. The adaptation process to frequent stimulus (e.g., training, learning) occurs in the following manner. There are two stages in response to an acute stimulus. In the first stage, in the presence of the acute stimulus the body tries to maintain its balance (e.g., consuming lactic acid generated during loading or converting it to pyruvate), and, depending on the intensity of the stimulus, the body

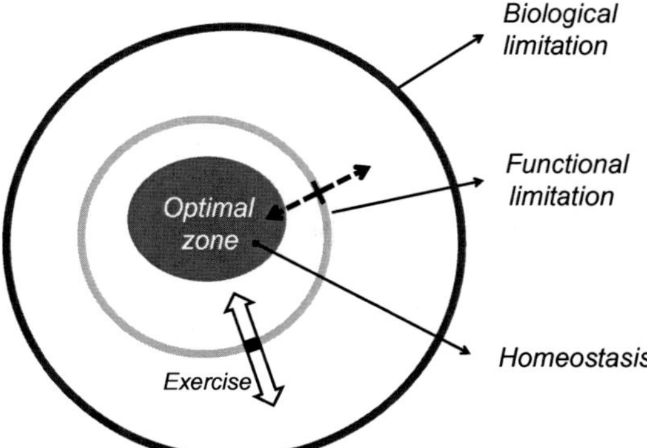

Fig. 3.4 Homeostasis and adaptation limits. Functional limits of untrained athletes stand far from their biological limitations, which can be approached by frequent physical training, thereby widening the zone in which the body can tolerate loading. Crossing the biological zone is not possible in this sense, because those conditions cannot be tolerated and cause death, like higher body temperature than 43°C.

may be able to preserve homeostasis, resulting in constant performance. However, continuous exposure to stimuli leads to decrease of internal reserves, imbalance of homeostasis (e.g., blood pH falls below 7.1), and decline in performance. In this stage, the body utilizes energy from anaerobic metabolism in addition to aerobic processes. On further exposure to stimuli, homeostasis cannot be maintained (e.g., blood pH falls below 6.8), resulting in fatigue (Fig. 3.4).

Imagine a circle where the outer circle defines the border of biological limits of an organism (e.g., blood pH 6.0, body temperature below 15°C or over 43°C). The optimal zone is located in the middle of this circle, and between functional and biological limitation there is a space where physical training is able to make a difference in an individual. Moderately trained or untrained athletes cannot get as far from the optimal zone as trained athletes after an episode of extreme training. Excellent endurance is reflected by a quick recovery and return into the optimal zone in the resting period (this will be considered in detail later).

The primary reaction to acute exercise can be interpreted as the exhaustion of the resources produced by physiological and biochemical reactions and psychological processes (e.g., motivation), resulting in impaired performance. This phenomenon is called fatigue. The development of fatigue depends on the characteristics of a stimulus/training. Continuous, low intensity training decreases the level of accumulated glycogen in the muscle, leaving lactic acid production low, whereas short, high intensity training (e.g., sprinting) does not affect the level of glycogen in the muscle, but increases the level of lactic acid significantly.

In the resting period following acute exercise, the second phase of adaptation allows the body to recover and return to the optimal zone. Trained athletes recover faster and more effectively compared to untrained ones.

According to Matveyev's periodization model, the body overcompensates in the resting period, resulting in an elevated baseline. The degree of overcompensation is higher in untrained younger individuals compared to older ones. Other factors that influence overcompensation include sport transfer mechanisms, genetics, and environmental factors. During frequent training, if amplitude of the stimulus and regeneration time are selected optimally, the rate of overcompensations can aggregate, and if overcompensation constantly occurs, it will result in improved performance, which is the goal of physical training.

The adaptation mechanism also involves gene expression; certain genes are activated as training starts, while others reach their peak several days following training, meaning that intramuscular changes occur for 1–2 weeks after exercising. Some adaptational changes occur even later, after several months of constant training, and some epigenetic alterations only occur after decades of frequent training. This will be discussed further in detail later, and specific adaptation, power, and endurance training will be addressed.

The characteristics of the stimulus determine the reaction to it, and that determines the factors of fatigue. These will be discussed in the next paragraphs.

3.2 FATIGUE

Several hours of reading causes the eyes and the brain to fatigue, which results in altered neural activity such as impaired concentration, attention, and memory. Frequent reading such as 1–2 h every day would delay the development of fatigue, resulting in improved performance, storing and remembering more information, and easier recall of this information afterwards.

Fatigue during a javelin throw training session manifests as impaired concentration and movement coordination, and impaired overall performance, which is caused by molecules that are involved in the development of fatigue such as Pi and ADP. In contrast, other factors are involved in the development of fatigue in a 1500 m swim. Therefore the intensity of the training determines the characteristics of fatigue, and regeneration is a response to the intensity-dependent fatigue. Thus, fatigue plays an important role in adaptation; it is another way that adaptation is determined by the intensity of the physical training.

We discuss fatigue based on the distinction of two main changes such as central nervous system-related mechanisms and musculature-related mechanisms. The former is a controlling system, while the latter is an executioner system. Central nervous system alteration involves adverse changes in the neuromuscular junctions; the decreased level of acetyl-choline results in impaired muscle contraction. Fatigue is an important defense mechanism of the body, therefore it needs to be considered seriously. Pain is often associated with fatigue, and is also a defense mechanism; therefore painkillers are sometimes included in doping lists, since these compounds disarm the alarm system of the body. Tolerance to pain can be increased within certain limits. Fatigue is also influenced by psychological factors. Patients suffering from chronic fatigue syndrome show the same pulse and lactic acid values compared to healthy individuals; however, they are more inclined to find training painful. Fatigue can be mild, moderate, or severe, and setting the optimal level of fatigue during training is difficult; however, it is an important factor in the adaptation process.

3.2.1 Central Fatigue and Nervous System Fatigue

Central fatigue occurs via regulatory responses of the nervous system to protect the body against injury. Blood flow to the brain increases by 25% during submaximal training; however, upon maximal loading it may decrease, as estimated by the elimination rate of radioactively labeled xenon.

Long-lasting endurance training or shorter periods at a high external temperature and humidity induce increased body temperature. Lack of water supply during physical training may result in a body temperature over 40°C, which risks the incidence of adverse physiological mechanisms. Increasing body temperature and volume contraction are associated during physical training, since sweating serves as the control mechanism to remove excess heat. Dehydration occurs if consumed body fluid is not compensated for by drinking plenty of fluids. During marathon running the weight loss can be 3–5 kg due to loss of fluids (sweating and evaporation). Compensation of loss of volume during training is a desirable habit that professional tennis players demonstrate during the breaks between tennis competitions. Drinking plenty of fluids delays the development of fatigue and prevents underperformance.

Long-lasting physical training results not only in decreased body temperature but also in decreased blood sugar level. Although the brain demands a high energy supply, it is not able to accumulate glucose, unlike the skeletal muscle and liver. The brain weighs about 2% of total body weight; however, it use 25% of total glucose consumption of the body. Upon physical training, the blood glucose level could decrease significantly. Hypoglycemia has adverse effects on the normal function of the brain, so it inhibits physical activity. Thus, carbohydrate consumption during competition (e.g., marathon, triathlon, tennis) delays the development of fatigue.

In addition to decreased blood glucose level, the level of certain amino acids such as isoleucine, leucine, and valine may decrease, leading to fatigue. Amino acid supply during a competition may ameliorate the rate of fatigue; therefore sport drinks may contain these compounds.

Fatigue is also associated with changes in the signal transduction processes in neurons such as increased tryptophan and serotonin, and decreased dopamine levels, which may be partially responsible for impaired locomotion. An increased tryptophan level is the focus of studies investigating the mechanism of central fatigue.

Strenuous physical training generates ammonia via the adenylate kinase reaction ($ADP + ADP = ATP + AMP$) from adenosine monophosphate, which cannot be further converted to ATP, therefore it is metabolized to ammonia. Ammonia is able to affect locomotion adversely since it influences the metabolic processes in astrocytes (star-shaped glial cells in the brain and spinal cord), and also affects neurotransmitters. Ammonia impairs locomotion, effective learning, and memory. In addition to ammonia, lactic acid also influences locomotion during high intensity training. Supplements that buffer lactic acid production slow down the progress of fatigue.

3.2.2 Peripheral Fatigue and Muscle Fatigue

Muscle fatigue is considered a natural consequence of physical training; however, its causes may differ depending on the intensity and duration of exercise. Low intensity long-lasting training decrease the glycogen content of the musculature. The muscular system, unlike the nervous system, is able to store huge amount of energy in the form of glycogen, which may decrease by 25%–30% during a 15 km flat race, as measured by muscle biopsy. After several hours of competition like a marathon or triathlon, decreased glycogen level may cause underperformance. Glycogen stored in muscle shows sporadic distribution in muscle fibers and also the differential activation rate of muscle fibers causes a diverse decrease, thus a 20%–30% decrease is manifest in different absolute levels in each muscle fibers. Decreased glycogen can be considerable during serial loading (e.g., European Championships,

World Championships, Olympic Games). Thus, carbohydrate supply before competition can prevent fatigue-related underperformance.

Skeletal muscle contains a high amount of nitrogen oxide synthase (NOS), which produces a relaxant, nitrogen oxide (NO). NO is also generated in smooth muscle cells (discussed in detail in the section on the cardiovascular system). Nitrogen oxide inhibits the attachment of actin–myosin filaments, and it has a role in the development of fatigue as well as after low intensity, long-lasting training in particular.

Muscle contractions are the results of a series of ion (Na^+, K^+, Ca^{++}, H^+) translocations during long-lasting training. The efflux of K^+ increases its concentration in the extracellular space, resulting in impaired muscle contraction. Calcium ions have long been associated with fatigue. Ca^{++} binding to tropomyosin results in uncovered myosin binding sites of actin. A lack of calcium ions prevents the development of cross bridges between actin and myosin fibers, leading to impaired contraction. This type of fatigue has been observed during low intensity, long-lasting training.

According to recent research, Ca^{++} from the endoplasmic reticulum stimulates mitochondrial ATP production in addition to its central role in muscle contraction. Over a certain level it increases the level of free radicals, which promotes the development of fatigue. Free radicals in low quantities stimulate contraction, whereas high amounts of oxidants inhibit it. The underlying mechanism behind stimulation involves their actions on signal transduction and results in the stimulated efflux of calcium ions from the sarcoplasmic reticulum, and increased sensitivity of calcium ion channels, which facilitates muscle contraction. High levels of oxidants, however, decrease calcium ion efflux from the sarcoplasmic reticulum, and also induce structural changes in proteins, resulting in impaired biochemical function and impaired muscle contraction. Thus, antioxidant supply either stimulates or attenuates the rate of muscle contraction.

As glycogen level decreases and promotes underperformance during long-lasting training, the decrease of CP (creatine phosphate) also promotes fatigue during high intensity training. Accumulation of creatine phosphate is limited; this mechanism will be discussed in the chapter about diet.

High intensity training promotes the increase of H^+, and lactic acid, which results in underperformance, while alkalosis aids in recovery. A low pH deteriorates the connection of actin–myosin muscle fibers. Actin–myosin binding and cross bridges formation involve a weak attachment state and a power stroke stage. A low pH increases the percentile of low attachment cross bridges, while that of power strokes decreases. In addition, a low pH during anaerobic conditions damages troponin structure, which directly impairs muscle contraction.

Impaired contraction can also be caused by increased levels of Pi and ADP; in addition, its role in fatigue has been explored in intensive muscle contractions. Pi and ADP compete with ATP for the ATP binding sites on myosin, and a high concentration of Pi or ADP deteriorates ATP binding and weakens muscle contraction. ADP-ATP translocation needs to be fast, which demands ATP, thus the level of Pi or ADP as side products of muscle contraction exceeds ATP in fast movements.

Calcium ion binding to tropomyosin is inevitable in muscle contraction. Impaired calcium ion reuptake by the sarcoplasmic reticulum prevents muscle relaxation, and the lack of relaxation of the antagonist muscle may result in a pulled or ruptured muscle. This can be seen in thigh flexors in sprinters during racing. Fatigue caused by impaired calcium ion reuptake occurs primarily during high intensity exercise; this can be attenuated by caffeine consumption, since caffeine promotes calcium ion uptake by sarcoplasmic reticulum.

In summary, the development of fatigue is influenced by the characteristics of physical training such as intensity, duration of the training, and also environmental factors (external temperature and humidity, audience, etc.). We need to emphasize that the adaptation profile and the time needed for regeneration depend on the characteristics of fatigue.

3.3 MUSCLE SORENESS AND OVERTRAINING

Both muscle soreness and overtraining result in underperformance. Both are the focus of research, and although a huge amount of data have been collected on them, the mechanisms are still not fully understood.

3.3.1 Muscle Soreness

Muscle soreness usually develops after unaccustomed, significant muscle contractions, primarily following eccentric and isometric muscle contractions. It is associated with pain and reaches its peak 24–48 h after training and results in underperformance and decreased maximal power generation of the muscle. The muscle structure contains several microscopic injuries. Sarcomeres resist contractions to various degrees, thus they suffer from damage to various extents. Stretches may cause such mechanical stress that myofilaments drift apart, and overlapping disappears, preventing the connection of actin and myosin, therefore preventing contraction. This damage usually involves other structural elements such as sarcolemma, Z-line and sarcoplasmic reticulum, and transverse tubules. Injury of the sarcoplasmic reticulum, which stores Ca^{++}, results in intracellular calcium ion increase leading to activation of proteases and degradation of the damaged proteins. The injury also activates the immune system, resulting in inflammation via cytokines, neutrophil granulocytes, macrophages, and lymphocytes. This process is an important part in healing.

It seems that IGF-1 (insulin-like growth factor 1) has an important role in regeneration, which is an anabolic hormone, and satellite cells activated at the site of injury contain high levels of IGF-1. During inflammation, the level of NO also increases, which can directly inhibit actin–myosin interaction, resulting in decreased power generation by the muscle and relaxation. It is also able to activate satellite cells (muscle stem cells), which first divide, then differentiate to muscle cells. Thus, NO has an important role in muscle regeneration, which, in addition to activation of satellite cells, stimulates follistatin production, enhancing regeneration (Fig. 3.5).

Nociceptors, pain receptors in muscle and connective tissue, send a stimulus to the spinal cord resulting in analgesia. These receptors can be

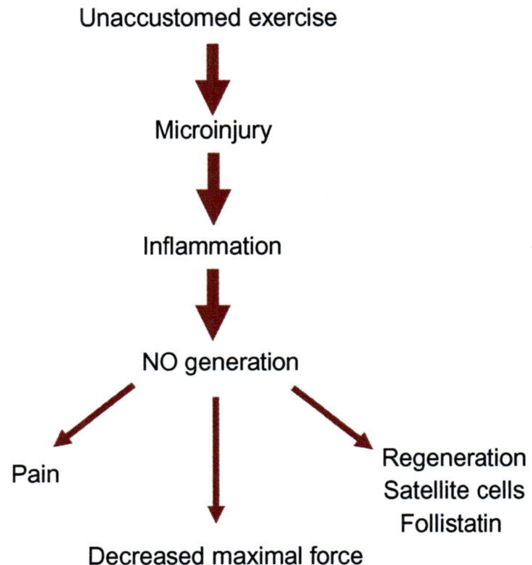

Fig. 3.5 Hypothetic model of muscle soreness. Unaccustomed, significant muscle contractions induce microscopic injuries in the muscle inducing inflammation with consequential nitrogen oxide (NO) production. NO production is also activated by other mechanisms during muscle contraction. NO decreases the maximal force generated by the muscle, which is a well-known phenomenon in muscle soreness. NO also stimulates division of satellite cells and follistatin, which are important in muscle regeneration. NO is able to induce pain, which is also a well-known phenomenon in muscle soreness.

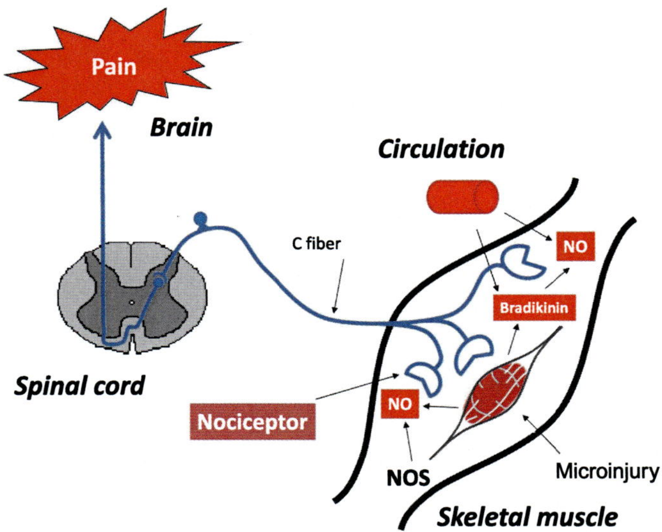

Fig. 3.6 Pain sensation during muscle soreness. Nociceptors in the muscle can be stimulated by NO and bradykinin. Type C nerve fibers transfers the signal to the spinal cord and to the brain.

activated by ATP, NO, glutamate, NGF (nerve growth factor), and bradykinin, whereas lactic acid is not an activator of nociceptors. Duration of pain is also determined by NGF, which stimulates unmyelinated type C nerve fibers in the muscle (Fig. 3.6).

Development of muscle soreness is a complex process, and several steps have not been explored yet. However, mechanical stress-related micro injuries and the consequential inflammation process are well-established. Lactic acid is not responsible for the characteristics of muscle soreness such as pain, microscopic injuries, and decreased muscle force generated by the muscle. Lactic acid's half-life is shorter than would be required to cause the symptoms of muscle soreness, which develops 24–48 h after training.

3.3.2 Overtraining

Overtraining also results in underperformance. It develops after a series of extremely high intensity trainings and is associated with other factors such as diet (e.g., weight loss diet), social (stress at school or work), and mental (e.g., relationship rejection) factors, making it a complex process.

Because of this complex process, there is no one special biological marker that could indicate overtraining. The main marker indicating overtraining is underperformance (Fig. 3.7); however, impaired performance may indicate other physiological disturbances as well.

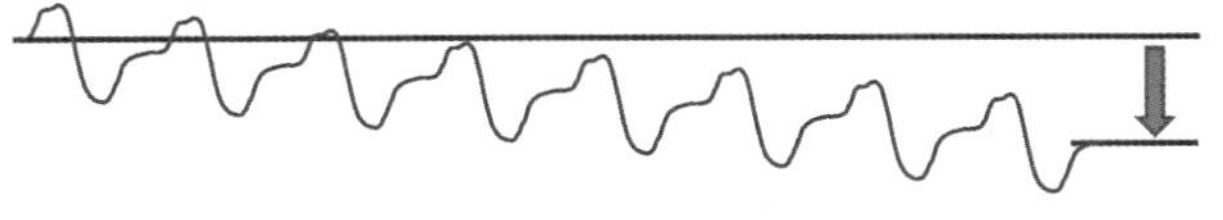

Fig. 3.7 Overtraining. Overtraining results in underperformance. It develops upon frequent, extremely high intensity trainings. The mechanism is a complex process, and important factors include miscalculation of loading intensity and necessary resting, emotional stress, faulty diet, and other bad habits.

Overtraining is used as a general term referring to the complex symptoms of impaired performance. Overtraining as a result of frequent physical training may become chronic if regeneration is not complete and fatigue is present over time. This may be a consequence of miscalculation of loading intensity and necessary resting time. Regardless, adaptation following physical training is specific and stimulus-dependent, thus there is a difference in the symptoms following short duration high intensity training and long duration low intensity training.

Overtraining can also be divided into sympathetic and parasympathetic categories. Symptoms of sympathetic overtraining indicate the effort of the body to maintain homeostasis, which can be seen as the first phase of overtraining. Because of the continuous loading, the body exhausts its stores of energy, accompanied by the dominance of parasympathetic regulation. This phase is similar to the stress reaction described by Selye. The last phase is exhaustion and low resistance to stress. There is a close association between muscle power and circulating adrenalin in trained athletes, which disappears upon overtraining indicating a neural dysregulation.

Scientific data on overtraining from athletes is limited and also data from laboratory animals is limited. Overtraining syndrome has physiological, biochemical, immunological, and psychological manifestations (Table 3.1). In practice, if an athlete underperforms and complains about chronic fatigue, resting heart rate needs to be measured and compared to his/her previous resting heart rate. In addition, the difference between heart rates in standing and lying position shrinks if overtraining occurs. The length of time needed to return to resting heart rate is also a useful indicator of overtraining; in overtrained athletes, more time is required.

Physiological characteristics of overtraining include the dominance of catabolic processes over anabolic resulting in weight loss, and loss of muscle. The rates of cortisol (a catabolic hormone) and testosterone (an anabolic hormone) in blood are used to estimate overtraining. However, there is no such test today that is guaranteed to indicate or predict overtraining.

Table 3.1 Physiological, biochemical, immunological, and psychological characteristics of overtraining

Symptoms of overtraining

Physiological	Biochemical	Immunological	Mental
Decreased performance	Increased level of protein breakdown	Increased sensitivity to all kind of diseases	Depression
Disrupt coordination			Decreased self-confidence
Decreased level of O_2 carrying capacity	Impaired insulin sensitivity	Impaired immune response	
	Increased catabolic process		
Impaired heart recovery	Impaired mineral absorption	Increased incidence of upper respiratory diseases	Emotional instability
Decreased training tolerance			
Chronic fatigue	Decreased muscle glycogen levels	Swollen lymph notes	Impaired focusing
Increased resting and delayed recovery heart rate	Increased levels of cortisol	Fever	
Decreased body mass	Decreased levels of testosterone	Increased incidence of bacterial infection	Fear
Decreased efficiency during exercise			Lack of endurability
Sleeping problems	Decreased glutamine levels	Increased appearance of herpes	
Sustained muscle and joint pain		Decreased number of lymphocytes	
Lack of appetite			

Biochemical characteristics include decreased glutamine level produced by the muscle, which also affects the efficiency of the immune responses. Thus, increased incidence of illnesses and increased length of time necessary to heal injuries are consequences of overtraining.

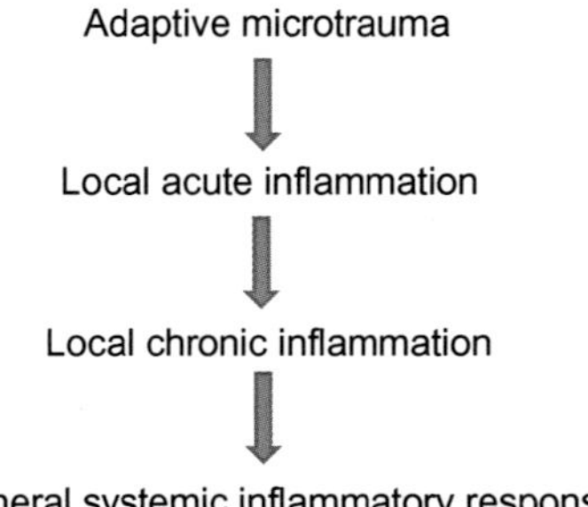

Fig. 3.8 The underlying theoretical mechanism of overtraining.

Short duration of overtraining does not affect recovery, and some trainers deliberately use overtraining as a prerequisite for future outstanding performance. This is very risky and also raises ethical concerns. Overtraining over time results in underperformance and puts an athlete's health at risk.

There are some theories that attempt to reveal the underlying mechanisms of overtraining. Apprehension of the mechanisms would provide the opportunity to prevent and treat overtraining effectively.

One theory focuses on microscopic injuries in the muscle that induce inflammation. Molecules such as cytokines, which can be produced in almost any cells, induce (proinflammatory) or reduce (antiinflammatory) inflammation (Fig. 3.8).

According to cytokine-overtraining theory, local cytokine production is aggravated, which leads to chronic inflammation. This theory is supported by the observation that exposure to cytokines in the body produced overtraining syndromes independent of physical training (Fig. 3.9).

Apprehension of the causes is important in conquering overtraining syndromes. If faulty diet or inadequate training are the causes, they can be corrected; if other factors such as health problem, social or psychological disturbance are the main causes, then athletes need to be supported to overcome underperformance. Physical loading needs to be decreased and its type changed as well. In addition, environmental changes, resting, a dietitian, and a sport psychologist can help.

3.4 REGENERATION AND RESTING

The main steps of the adaptation process in response to physical training occur in the resting phase. The length of the resting period and its characteristics determine the efficiency of the adaptation. In professional sport, careful design of the resting periods between multiple trainings on a single

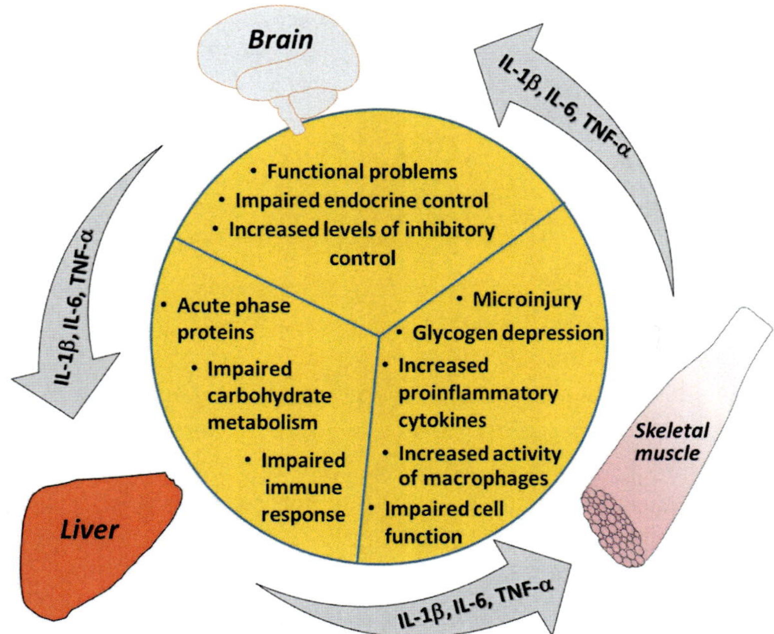

Fig. 3.9 Regulatory functions of cytokines in brain, liver, and muscle functions. Cytokines are able to induce functional changes in the brain, liver, and muscle, resulting in overtraining syndromes.

day is vital. Regeneration during the resting periods can be accelerated by physiotherapy, massage, muscle cooling, hot and cold therapy, ultrasound, magnetic therapy, and electrolyte infusion.

Different types of fatigue require different durations of regeneration. Anaerobic-alactic-related fatigue require short regeneration times—for instance, fatigue following a 100 m running qualifying round in the Olympics can be overcome quickly. Anaerobic-lactic-related fatigue, where lactic acid concentration may reach 20 mmol/L, needs more time. During competitions, lactic acid returns to its base level after 30–40 min, e.g., allowing athletes to be ready for the tandem kayak race 1 h after a single kayak race.

The accumulation of side products is primarily responsible for anaerobic fatigue. Aerobic fatigue, however, requires a longer regeneration period; for instance, a 10,000 m run cannot be repeated each day in a competition because of the depletion of glycogen stores. In this case regeneration requires a longer time. Characteristics of different types of training are listed in Table 3.2.

Table 3.2 Characteristics of fatigue and regeneration following different training intensity

Duration of loading	Examples	VO$_2$max%	Metabolic contributions (estimation)			Regeneration time
			Anaerobic alactic	Anaerobic lactic	Aerobic	
3–20 s	Ball game dash, 200 m dash	>160%	~15%	~60%–65%	~25%–30%	Some min
20–60 s	400 m dash, 50 m swim	~140%–150%	~10%	~45%–60%	~30%–45%	Some min
90–120 s	800 m dash, 500 m kayak, 200 m swim	~115%–130%	~5%	~30%–45%	~50%–65%	30 min–1 h
3–5 min	1500 m run, 400 m swim, 1000 m kayak	~105%–115%	~2%	~15%–28%	~70%–85%	30 min–1 h
5–9 min	800 m swim	~90%–100%	<1%	~8%–12%	~85%–90%	Several hours to 1 day
20–90 min	10,000 m run, ball games	~60%–90%	<1%	~5%–15%	~85%–95%	3–6 days

Sleeping is one of the most important parts of the regeneration process. Its efficiency is determined by the timing of training and competition; vice versa, sleeping affects performance during subsequent training and competition. In athletes who suffer from sleeping disorders, aerobic training improves sleeping efficiency (Reid et al., 2010). Taking a nap after a training session stimulates regeneration efficiency (Davies et al., 2010). Late trainings or competitions affect deep sleep and regeneration efficiency. Altitude or jet lag after a long flight can also cause sleeping disturbances, which alter performance.

3.5 PRINCIPALS OF EXERCISE TRAINING

Since the purpose of this book is to discuss the sport physiology of conditional abilities, strength, endurance, speed, and flexibility (the latter also belongs to coordination skills), we need to address the fundamentals of exercise training.

Conditional abilities are in strong association with each other, therefore any separations are acceptable only for didactic purposes. For instance, strength generation is always essential to initiate locomotion; a 100 m run may last for only a few seconds but it require endurance as well; and to win an Iron Man triathlon requires speed/velocity in addition to endurance. We attempt to separate and characterize conditional abilities and shed light on the most important differences and similarities by considering those factors that have importance in these skills.

Strength is primarily determined by musculature, the hormonal and nervous systems, and genetic factors. Endurance is determined by the cardiovascular system, muscular system, psychological factors such as stress tolerance, and genetic factors. Speed is determined by the nervous system, musculature, and genetic factors. Flexibility is determined by musculature, tendons, and genetic factors.

Accomplishment of optimal adaptation needs carefully designed physical training of optimal intensity and duration. Intensity and duration are inversely proportional, since maximal speed only can be maintained for a short time, and also low intensity movements can be maintained for longer.

Intensity can be set by individual maximal oxygen uptake (VO_2max) or group VO_2max in a team. Endurance is determined by the relative VO_2max percentage; alternatively, heart rate can be used if it is monitored continuously during training. Strength is determined by measuring individual maximal value as 100% and setting the training intensity according to that value.

Flexibility can be determined by, for example, the maximal movement in a given joint. Intensity is influenced by external factors; for example, maximal intensity can be different during training and during a local competition, national competition, or at the Olympics. These factors are addressed in detail in several books about training fundamentals. Duration of loading can be set in time (e.g., 1 h running), distance (e.g., 4 km swimming), and serials and repetition (e.g., 5 × 20 sit-ups).

3.6 ACCLIMATIZATION

Long-distance travel is a common occurrence in order to participate in a competition in the modern world. Jet lag and humidity and temperature differences can be a challenge to the body, influencing its performance. The body is influenced by the cyclic changes of light and darkness, called the circadian rhythm. Physical changes include core body temperature as well, with the highest value around 4–8 p.m., and the lowest around 3–6 a.m. The "internal clock" is localized in the suprachiasmatic nucleus of the hypothalamus, which regulates body temperature and hormonal secretion. This clock is set by the retinal sensation and consequent melatonin production by the pineal gland. Melatonin is the main regulator of the circadian rhythm (Fig. 3.10).

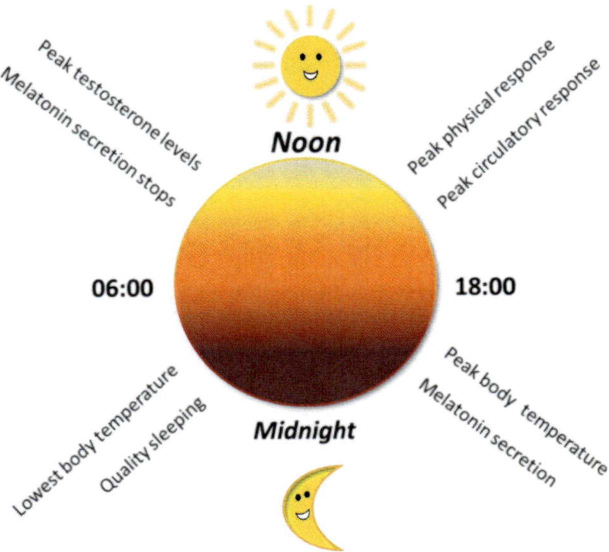

Fig. 3.10 Physiological changes in circadian rhythm.

During the circadian cycle, activity of the immune system is the lowest around 7 a.m., while mental functions reach their peak around noon, which correlates with adrenalin levels. Physical peak performances occur around 4–8 p.m. Daily differences in peak physical performance can reach 2%–11%, indicating the importance of the circadian rhythm in sport. It has been noticed in American football competitions that underperformance occurred when the tournament appointed time conceded with the midnight of the visiting team's circadian rhythm. Individual differences, however, can be significant; some athletes are not affected by passing time zones.

Melatonin regulates the daily cycle of body temperature and antiinflammatory cytokine production; it also has an antioxidant effect. Thus melatonin intake ameliorates circadian rhythm-related performance disturbances. Melatonin intake, following consultation with a doctor, is useful at night or during night flight, and in the morning 460–480 nm blue light helps reduce melatonin production, resulting in a quick adaptation and reducing jet lag symptoms. If a competition is due to take place 3 days after long-distance travel (e.g., a 1-day competition in the Far East), the athlete's daily rhythm should be arranged around local time. If they will be staying for longer than 3 days, then adaptation is more profitable, for instance, a morning exercise and avoiding evening trainings will help ameliorate jet lag symptoms. Since the circadian rhythm affects the secretion of several hormones that influence physical performance (e.g., growth hormone), quick adaptation prevents underperformance. It is important to note that psychological factors also determine performance after long-distance travel. Jet lag occurs when someone quickly travels across multiple time zones, and the inner clock is still synced to the original time zone, resulting in temporary sleep disorder, fatigue, and underperformance. The same phenomenon can be noticed with shift workers since there is a disharmony between the inner clock and the day/night cycle.

3.7 "R" EXTRA

According to the legend, Diotimos's son, Milo from Croton, entered for seven Olympics. He had huge strength owing to his training plan, which involved lifting a heifer from its young age until it had grown. It happened once that he took a heifer with him to a competition, where he trained by lifting the animal, then ate its meat, thereby also contributing to his outstanding performance.

Hormesis theory proposes that biological systems respond with a bell-shaped curve to stressors (Fig. 3.11). This indicates the existence of optimal

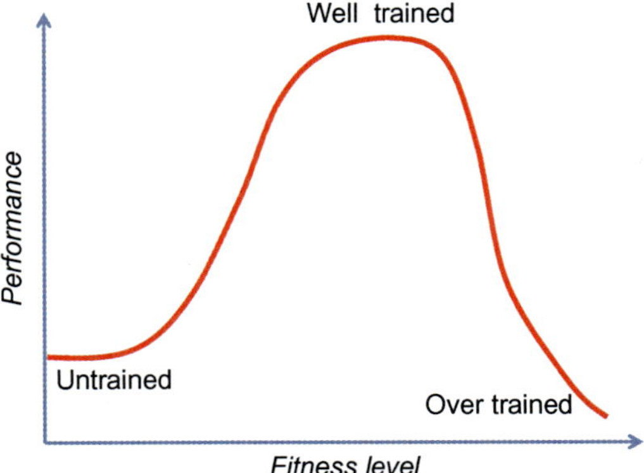

Fig. 3.11 Hormesis theory of physical fitness. The correlation between physical fitness and performance forms a bell-shaped curve. This curve is true for several phenomena.

zones, and deviations from optimal values result in adverse responses. This is probably true for physical training, where low values of the curve indicate physical inactivity, and high values indicate overtraining. Optimal loading results in long-lasting adaptation with beneficial effects such as longevity, decreased incidences of morbidity and mortality, and lifestyle-related diseases such as type II diabetes, cardiovascular disorders, stroke, and breast, prostate, and colon cancer. Physical training has only beneficial effects, except in the case of overtraining. Professional athletes live longer compared to the average population; however, some adverse effects related to the skeletal system can be observed because of the extremely high loading.

The degree of adaptation decreases with age, and it is significant after the age of 60, but can be controlled by frequent physical exercising.

3.8 SUMMARY

Adaptation is the body's reaction to external or internal changes. It has physiological and biological limits; however, its functional limits can be broadened. Its quality is stimulus-dependent, it manifests in multiple steps and level including molecular level, cellular level, and at the levels of organs, and at the phenotypic level of the entire body. A stimulus needs to be adequate with optimal intensity to induce adaptation, and results in impaired performance because of fatigue. Fatigue depends on the intensity and duration of the training, and also on resting. An apprehension of the characteristics of central and peripheral fatigue provides an opportunity to control its development and enhance physical performance. Muscle soreness is caused by microscopic injuries and results in pain. Overtraining is a complex process determined by multiple factors, which may result in long-lasting under-performance. Symptoms of overtraining can be divided into physiological, biochemical, immunological, and psychological groups.

TEST QUESTIONS

1. How do homeostasis and adaptation correlate?
2. What are the main types and characteristics of adaptation?
3. What are the causes of central and peripheral fatigue?
4. What are the characteristics of muscle fever?
5. What are the symptoms of overtraining?
6. What opportunities does a trainer have to prevent overtraining and to enhance regeneration?

BIBLIOGRAPHY

Borresen, J., Lambert, M.I., 2009. The quantification of training load, the training response and the effect on performance. Sports Med. 39, 779–795.

Davies, D.J., Graham, K.S., Chow, C.M., 2010. The effect of prior endurance training on nap sleep patterns. Int. J. Sports Physiol. Perform. 5, 87–97.

Flueck, M., 2009. Plasticity of the muscle proteome to exercise at altitude. High Alt. Med. Biol. 10, 183–193.

Issurin, V.B., 2010. New horizons for the methodology and physiology of training periodization. Sports Med. 40, 189–206.

Kjaer, M., Langberg, H., Heinemeier, K., Bayer, M.L., Hansen, M., Holm, L., Doessing, S., Kongsgaard, M., Krogsgaard, M.R., Magnusson, S.P., 2009. From mechanical loading to collagen synthesis, structural changes and function in human tendon. Scand. J. Med. Sci. Sports 19, 500–510.

Lundby, C., Calbet, J.A., Robach, P., 2009. The response of human skeletal muscle tissue to hypoxia. Cell. Mol. Life Sci. 66, 3615–3623.

Morton, J.P., Kayani, A.C., McArdle, A., Drust, B., 2009. The exercise-induced stress response of skeletal muscle, with specific emphasis on humans. Sports Med. 39, 643–662.

Murase, S., Terazawa, E., Queme, F., Ota, H., Matsuda, T., Hirate, K., Kozaki, Y., Katanosaka, K., Taguchi, T., Urai, H., Mizumura, K., 2010. Bradykinin and nerve growth factor play pivotal roles in muscular mechanical hyperalgesia after exercise (delayed-onset muscle soreness). J. Neurosci. 30, 3752–3761.

Murton, A.J., Greenhaff, P.L., 2010. Physiological control of muscle mass in humans during resistance exercise, disuse and rehabilitation. Curr. Opin. Clin. Nutr. Metab. Care 13, 249–254.

Radak, Z., Chung, H.Y., Koltai, E., Taylor, A.W., Goto, S., 2008. Exercise, oxidative stress and hormesis. Ageing Res. Rev. 7, 34–42.

Reid, K.J., Baron, K.G., Lu, B., Naylor, E., Wolfe, L., Zee, P.C., 2010. Aerobic exercise improves self-reported sleep and quality of life in older adults with insomnia. Sleep Med. 11, 934–940.

CHAPTER 4

Fundamentals of Strength Training

Strength is essential in any sport. Different sports require different types and degrees of strength. For instance, the apparent results of strength training may show differences such as when comparing Arnold Schwarzenegger, who had huge biceps, and Bruce Lee, who had lean muscles but was able to break many bricks with a single strike. The heaviest weight a man could lift over his head is over 260 kg. This nicely shows the broad range of adaptation following strength training. This chapter addresses the physiological and methodological features of strength training.

Types of strength include maximal strength, strength endurance, explosive strength, and plyometric strength. Relative strength is the amount of force generated per unit of bodyweight. Rational and sport-specific design of strength training is a complex task because of the different attributes of strength. The force generated by the muscle is influenced by several factors at micro and macro levels:

- Cross bridges formed in a sarcomere over a given time determine the force that the muscle can generate.
- The dominant types of muscle fibers determine the type of sport in which an athlete can be successful; for instance, fast-twitch muscle fibers generate greater force, thereby they have special importance in power sports.
- The amount of motor units active at a given time also determines the force that the muscle can generate.

The synchronicity, synchronological, or consecutive activation of muscle units is also a determinant of force generation. Synchronization is an important enhancer of muscle force generation.

- Maximal force generation is determined by the characteristics of muscle contraction. In concentric contraction, maximal force generation is the lowest, compared to isometric, and it is greatest in eccentric contraction.
- The arrangement of muscle fibers, whether they are parallel muscles or pennate fibers, also determines maximal force generation.

The Physiology of Physical Training
https://doi.org/10.1016/B978-0-12-815137-2.00004-8

- There is a linear correlation between cross-sectional area or muscle hypertrophy and force generation; for instance, the quadriceps femoris generate greater force than the biceps brachii. This observation is important and indicates that one possible way to improve strength is through increasing the cross-sectional area of the muscles. This is regulated strictly by hormonal factors, so these concepts will be discussed together.
- The levels of anabolic hormones produced by the endocrine system has special importance for musculature; thus the age and sex of an athlete determine the force that a muscle can generate.

In the next paragraphs, these factors will be discussed in detail.

4.1 FORCE GENERATION AND ADAPTATION IN SARCOMERES

Unlike industrial simple motoric units, muscles are comprised of thousands of microscopic units called sarcomeres. They are arranged in a strict geometric network inside the muscle fibers and their parallel arrangement allows the generation of an enormous Newtonian force. Muscle contraction is the result of a series of molecular changes including efflux of Ca^{2+} ion from sarcoplasmic reticulum and its binding to troponin C, and consequent conformational change of the latter, which allows actin-myosin cross bridge formation. The length of a sarcomere influences Ca^{2+} sensitivity of the filaments, resulting in different numbers of cross bridges being formed, and consequentially affecting the force that the muscle can generate. The greatest force is generated during the power stroke phase of muscle contraction. In this phase, the head of the myosin filament is attached to actin for about 2 ms and generates 3–4 piconewtonian force. Imagine how many cross bridges need to be formed to lift 100 kg of weight.

Overlapping between myofilaments also influences force generation. If actin and myosin greatly overlap, there is only limited space for shortening, thus tension generation is reduced. Experimental data from frog muscles show that on decreasing sarcomere length from 2.1 to 1.7 µm, the maximal tension drops to almost zero (McDonald et al., 1997). Similarly, if overlapping covers a very small area because of stretching, it also decreases the extent of cross bridge formation between actin and myosin filaments and consequentially there is decreased force generation. In a sarcomere, there is a reverse correlation between the velocity of contraction and force generation. Maximal velocity can be reached at zero resistance, whereas at maximal resistance the muscle cannot shorten further, thus tension generation is isometric. The first phase of adaptation following strength training manifests in better innervation, improving neuromuscular coordination, recruitment of more motor units, and their synchronistic activation.

Muscle cell adaptation results either in hypertrophy or hyperplasia. Muscle hypertrophy is an increase in size of muscle fibers, whereas hyperplasia is an increase in the number of muscle fibers. The latter will be discussed further in the "R" Extra section at the end of this chapter. An apprehension of the molecular mechanism of hypertrophy and hyperplasia helps when selecting the optimal stimulus, its intensity and duration in a strength training session.

4.2 TYPES OF MUSCLE CONTRACTION AND FORCE GENERATION

Force generation in different types of muscle contraction such as concentric, isometric, and eccentric were discussed in Chapter 2, and recruitment of contractile element, series, and parallel elastic elements determine the degree of force generated by the muscle. It should be noted that in concentric and isometric contraction, recruitment of motor units is deliberately regulated, whereas in eccentric contraction there is an additional automatic activation since the external force on the muscle is greater than the force that the muscle can generate, thus defense mechanisms activate reflex-movements as well, and these together provide a higher force. Calculating the difference between maximal force generated by eccentric and isometric contraction raises the opportunity to design an optimal training program. Measurement of maximal force generated by the muscle can only be implemented in a laboratory, which is affordable in the case of professional athletes.

When the maximal force generated by the muscle during eccentric and isometric contraction is equal, it indicates that the cross-sectional area of the muscle is maximally exploited by the athlete; thus improving performance is possible via training that induces muscle hypertrophy. When there is a significant difference between maximal force generated during eccentric and isometric contraction, it indicates insufficient exploitation of the existing mass of muscle, thus training design needs to be focused on exercises that improve neuromuscular coordination to recruit more motor units and synchronize their activation.

4.3 MOTOR UNITS AND FORCE GENERATION

There is a significant difference between slow-twitch and fast-twitch muscle fibers, as discussed in Chapter 2, in their activation threshold and force generation. A question arises regarding why motor units need to be distinguished from the whole muscle. Unlike in cardiac muscle where all the cardiac cells are working as one syncytium and there are no motor units, in skeletal muscle, motor units (muscle fibers innervated by a single neuron) are recruited and activated. Force generation by skeletal muscle is several hundredfold greater than by that of cardiac muscle, which requires complex innervation. Recruitment and synchronicity of motor units can be measured by electromyography (EMG); however, the types of motor units can only be determined indirectly based on physiological characteristics such as fatigue and extent of force. As mentioned before, characteristics of muscle depend on innervation, which can be modified to a limited extent. Recruitment of motor units starts with the small ones, which generate the least force, and is followed by motor units that are greater in size and generate a significant force (Zajac and Faden, 1985). This indicates that recruitment of motor units aligns with power necessary for a movement; for instance, no recruitment of large motor units occurs if 5% force is necessary for a movement assuring profitability. When huge power is needed, in addition to fatigue-resistant motor units, other fatigable motor units with higher activation thresholds are recruited. Thus, force generation increases with the recruitment of motor units in a nonlinear manner since force generation by the different units is not equal. The correlation is exponential since motor units with the greatest force generation ability are recruited last (Fig. 4.1).

Force generated by different muscle fibers differs significantly even if a muscle, e.g., the soleus, contains mainly slow-twitch fibers; force generation falls between 31 and 1600 mN (McDonagh et al., 1980). This difference can be extreme in the case of a muscle comprised of mainly fast-twitch fibers, e.g., the tibialis anterior. It should be noted that force generation can differ between the same types of motor fibers, which can be monitored by EMG, where fiber types are reflected by activation frequency. The abovementioned activation order can be seen in slow movements, whereas in fast movements task-dependent recruitment has been observed. In latter case, e.g., learned movements, fast motor units can be recruited earlier regardless of normal activation order (Herrmann and

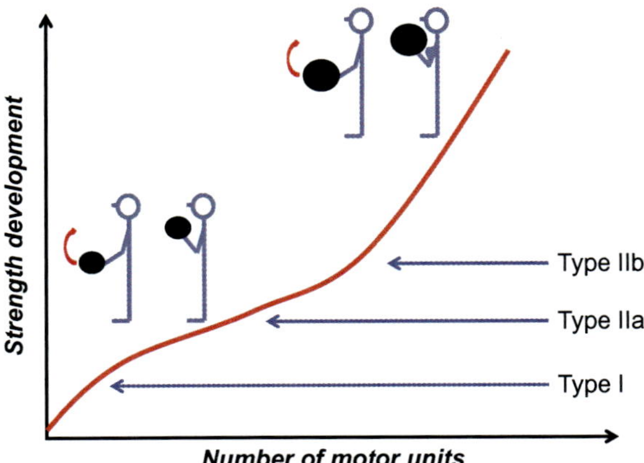

Fig. 4.1 Correlation between loading and recruitment of motor units. Slow profitable motor units (type I) are recruited first because of their low threshold. If greater force is needed, more and more intermediate (type IIa) motor units are engaged, then finally fast motor units (type IIb) with high activation thresholds are also enrolled. Slow fibers generate the least force, whereas fast fibers generate the greatest force.

Flanders, 1998). This observation emphasizes the potential usefulness of smooth automatic movements in explosive strength training. This will be further discussed at explosive strength training.

4.4 SYNCHRONIZATION OF MOTOR UNITS AND FORCE GENERATION

Force generated by each motor unit can be aggregated by synchronization of motor units that is their activation at the same time. Recruitment of motor units starts with the small ones and is followed by those that are larger, resulting in exponential force generation. A muscle comprised of heterogeneous motor units (motor unit pool) incorporates smaller and bigger motor neurons. When all the motor units are activated, this results in a contraction at 60%–80% of possible maximal contraction depending on the type of muscle. The additional force (20%–40%) comes from the synchronization of motor units (Duchateau et al., 2006). Although recruitment of motor units is based on their size described above, a particular load cannot be defined where a given motor unit is surely activated since activation is determined by the type of contraction (concentric, isometric, eccentric) and velocity. Synchronization of motor units examined by electrode needle placed in the muscle showed increased synchronization in trained college students and greater force generation, reflected by experimental data at 30% and 80% of maximal force generation (Fling et al., 2009). Thus, synchronization of motor units can be improved by physical training.

Investigating the effect of physical training in the elderly showed improved strength with low activation frequency of the motor units, which indicates improved synchronization of the motor units (Maejima et al., 2007). Thus, the body learns during physical training how to exploit the muscles more effectively resulting in more profitable functioning, which is necessary in endurance sports. Activation threshold of motor units decreases in contractions at high velocity. In this case, three times as many motor units are activated compared to muscle contractions at low velocity where the force generated by the muscle is the same. At 30% of the maximal force, all the motor units can be activated. The force generated by the muscle can be improved by physical training depending on the individual's ability to learn to synchronize and exploit the motor units efficiently. Adaptation to physical training includes agonistic and antagonistic muscle coactivation. This coactivation depends on the type of contraction, velocity, and fatigue. Advantages of the adaptation include joint stability and decreased antagonist exertion, resulting in profitable and fast locomotion. Activation and synchronization of motor units will be discussed as neuromuscular coordination, which has an importance in sports, especially sports requiring power and speed. The mechanism behind improved power in untrained individuals involves factors such as coactivation and improved synchronization of motor units (Fig. 4.2). When new motor units are recruited in muscle contraction they need to be harmonized with the other motor units, and synchronize their activation frequency for optimal locomotion.

4.5 MUSCLE HYPERTROPHY

Increased cross-section area (size) of muscles can be important for bodybuilders in particular, as well as other athletes, and also for the elderly to prevent diabetes. It has long been known that muscle strength is relative to cross-sectional area, which is reflected by different muscles and their relative strength. Thus, to improve power one needs to increase the size of the muscles. However, in some sports huge muscles can be a disadvantage. It is important to analyze the characteristics of a given sport when considering power training. For instance, in water polo, American football, and European handball, bigger athletes have advantages over smaller competitors. In contrast, volleyball players do not benefit from big size, which can

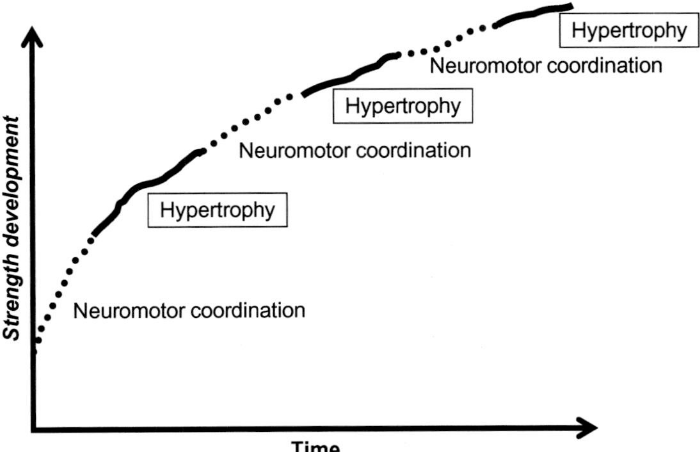

Fig. 4.2 Assumed mechanism behind improved power. During power training, the mechanism behind improved power involves factors such as coactivation and improved synchronization of motor units in the first phase, which is followed by muscle hypertrophy. This is followed again by improved synchronization. The rate of advancement shrinks in excellent power conditions since the potential to improve synchronization and hypertrophy further is limited.

impair overall performance. To increase muscle size, hormonal factors are essential in addition to mechanical factors. Three main anabolic hormones, growth factor, testosterone, and insulin-like growth factor (IGF-1), have importance in muscle growth. Muscle growth occurs when anabolic processes overbalance catabolic actions. Metabolic processes are complex and it is difficult to distinguish between the role of hormonal and mechanical factors in muscle growth (Fig. 4.3).

4.5.1 Mechanical Factors

Muscle fibers/cells differ from an average cell since they have multiple nuclei, and this allows them to grow in size. The first articles about myoblast fusion as the mechanism of multinuclear muscle fiber formation were published in the 1960s. Thus, multinuclear characteristics of muscle fibers develop during differentiation. In order to grow further in size, muscle require new nuclei since the capacity of a single nucleus to produce muscle fibers is limited. In a muscle cell the total surface area of organelles can be five times bigger than in any other cell type. Anabolic enzymes and steroids are able to increase the number of nuclei, resulting in efficient protein synthesis. Only one question remains: from where do new nuclei arise? Inactive (quiescent)

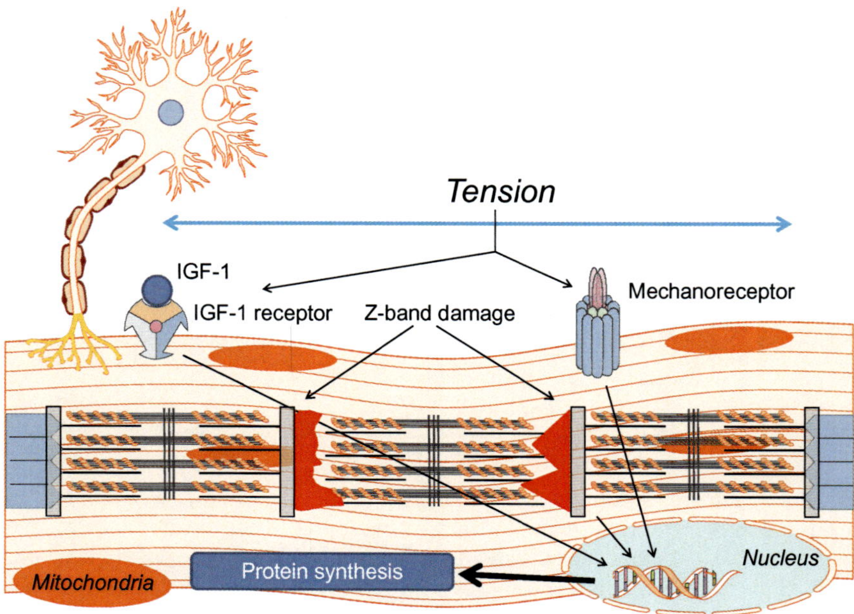

Fig. 4.3 Molecular mechanism of muscle growth. The increased cross-section area (size) of muscles is determined primarily by mechanical tension and the level of anabolic hormones (growth factor, testosterone, IGF-1). The levels of these hormones are influenced by the intensity of the training. Injury in the sarcomeres at the *Z*-line can be an initiation step for protein synthesis. Fast-twitch muscle fibers contain thinner *Z*-lines compared to slow-twitch fibers, therefore their sensitivity to mechanical tension and muscle growth is greater.

cells were noticed between the muscle fibers under the basal lamina, and these were called satellite cells because of their position. These cells are able to be activated and divided, and deliver their nuclei to muscle cells, resulting in an increased number of nuclei in the muscle fiber. Activated satellite cells are also able to differentiate to myoblast via activation of MyoD, a myogenic regulatory factor (Fig. 4.4). According to current knowledge, satellite cells are important and may be necessary components of muscle growth following power training. It is an interesting observation that if physical training is interrupted (e.g., due to injury), the newly integrated nuclei remain in the muscle for 3 months, thus muscle can grow quickly and effectively after restarting. These cells represents a kind of "muscle memory" (Bruusgaard et al., 2010).

Tension in the muscle, especially in the sarcomere, greatly influences the characteristics of muscle adaptation. Satellite cells can be activated by notable tension, which is important in cross-sectional area growth. Microscopic

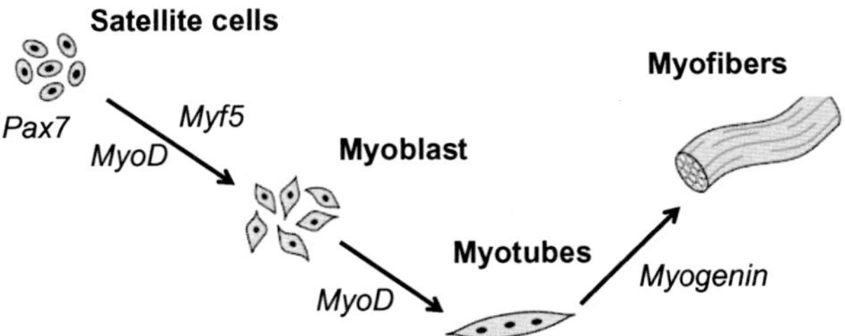

Fig. 4.4 Development of skeletal muscle. Development of skeletal muscle has several stages, and satellite cells are important actors in the process. Division and differentiation of these cells to multinuclear myoblast is an important step in development. Skeletal muscle is the only tissue comprised of multinuclear cells in the human body. The number of nuclei in a muscle cell may play an important role in adaptation of the muscle.

injuries in the muscle (e.g., Z-line stretching or rupture) caused by duration (serials and repetitions, length of isometric contraction) and intensity induce the reparation processes. In the extracellular matrix outside the sarcomeres there is a growth factor, hepatocyte growth factor (HGF), which is capable of activating the satellite cells. Microscopic injuries caused by tension are important stimulators of the HGF-satellite cell system. Upon tension permeability of ion channels are altered, which also could initiate protein synthesis in the muscle, resulting in muscle growth.

IGF-1 is also an important key factor in cross-sectional area growth. In addition to circulating IGF-1, muscular IGF-1, mechano growth factor (MGF), is also an inducer of protein synthesis in the muscle. An activator of MGF is mechanical tension and it has a threshold; thus tension needs to reach a certain intensity to be effective.

Dystrophin associated nNOS can also be activated by tension; it can stimulate satellite cells, and also influence directly the quantity of contractile proteins such as actin and myosin heavy chains (MHC).

During the reparation process following injury, macrophages (a type of white blood cells that engulf and digest debris) are also activated and produce IGF, which has an anabolic effect, thereby playing a role in the reparation process, and in supercompensation (Lu et al., 2011). Training with light weights and insignificant resistance does not activate these sensors and do not cause microscopic injuries, thus no significant protein synthesis occurs in the muscle.

Bodybuilders' quadriceps after 2 years of frequent training (with 60%–80% of single repetition maximal weight, 6–8 repetitions, and 4–6 serial) were 50% bigger compared to control (D'Antona et al., 2006). MHC can serve as a marker to distinguish between different types of muscle fibers. Fast-twitch type MHC increased more compared to slow-twitch type MHC in bodybuilders, regardless of the speed at which they executed their exercises. This would suggest that adaptation were not specific since locomotion were rather slow; however, fast-twitch muscle fibers are more vulnerable because of the weaker connective tissue, therefore they are more capable of hypertrophy. Based on the abovementioned observations, duration and intensity of the muscle tension are the most important factors behind muscle growth. The cross-sectional area growth is more robust in males than in females or children, suggesting that hormonal factors also must be considered.

4.5.2 Hormonal Factors

In this part, the effects of three main anabolic hormones—growth hormone (GH), IGF-1, and testosterone—on muscle hypertrophy are discussed.

Growth Hormone

This is produced in the anterior lobe of the pituitary gland or hypophysis, and it reaches the target organs via the bloodstream. Its main effects include the stimulation of muscle growth and fat metabolism, increasing bone density, improving immune responses, and sexual activity. Its level shows daily fluctuation, with the highest point during the night, and the daily fluctuation level is 5–40 ng/L; with age it decreases, with a peak level during youth. GH is used in anti-aging therapy and also in professional sports, since it has a power amplifier effect. It could not be detected from urine until the early years of the 21st century, but recently GH from outside sources can be traced both from the urine and the blood in doping tests. There are adverse side effects of long-lasting administration at high doses including diabetes and joint pain, and the incidence of cancer increases.

GH targets organs including the muscle and the liver, where it stimulates IGF-1 production. Upon high intensity physical training, GH level increases after 10–20 min. It influences metabolic processes and thus has positive effects on endurance markers such as maximal oxygen consumption (VO_2max). Its effect on muscle results in a cross-sectional area increase, but effects of GH and IGF-1 are closely associated (Fig. 4.5). Arginine intake stimulates GH production (Collier et al., 2005), therefore arginine is a highly preferred anabolic by bodybuilders. For females, GH-IGF-1 is important since the effects of testosterone are marginal.

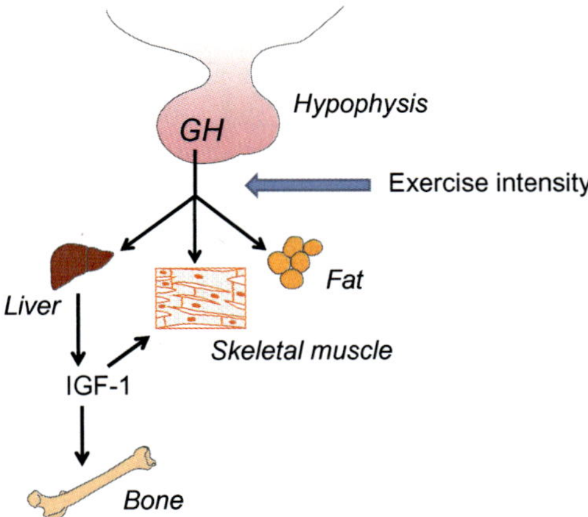

Fig. 4.5 Effects of growth hormone. Anabolic hormones are important in cross-sectional muscle growth, and their production is influenced by the intensity of physical training. Growth hormone stimulates IGF-1 production by the liver, and GH has a direct effect on the muscle, and also alters metabolic processes. IGF-1 also directly affects the muscle and metabolic processes.

IGF-1

IGF–1 is an anabolic hormone having similar structure to insulin, and it is produced by the liver; production may also occur in muscle locally having a more effective impact (Frystyk, 2010). Daily fluctuation of IGF–1 is not significant compared to GH or testosterone. Its concentration in the blood is 10–1000 ng/L depending on age, sex, genetic factors, and physical fitness. IGF–1, like many hormones, operates via receptors. Its main function is to stimulate protein synthesis. Since IGF–1 is a growth stimulator, its chronic high level increases the incidence of different type of cancers, and a low level of IGF–1/insulin promotes longevity. Long–lasting aerobic training decreases IGF–1 level, while high intensity training increases it (Kostka et al., 2003; Nishida et al., 2010). The level of IGF–1 increases upon muscle injury, and its anabolic effect promotes the healing and adaptation processes. Since IGF–1 level is closely associated with GH production, its level declines by age, similar to GH.

Testosterone

This is an anabolic steroid, produced mainly by the testis, and in a low extent by the adrenal gland. Slow– and fast–twitch muscle fibers show differences in sensitivity to testosterone. Interestingly, slow type I fibers shows higher sensitivity, while fast IIb are less sensitive to testosterone, regardless of the observation that cross–sectional area composed of fast IIb fibers are bigger than slow type I fibers in bodybuilders and weight-lifters. This phenotype is the result of heavy weight training. It is interesting to note that testosterone do not affect the transport of amino acids into the muscle; however, it promotes the metabolism of amino acids already present and their conversion to proteins. Skeletal muscle cells have a multinuclear structure, and the number of the nuclei may increase during power training, therefore they synthesize high amount of protein. Protein synthesis is promoted by the Akt and mTOR pathway, which can be stimulated by high muscle tension, IGF-1, and testosterone (Fig. 4.6).

The number of nuclei in muscle may increase in response to testosterone. Since muscle cells are not able to divide, satellite cells under the basal lamina differentiate and deliver their nuclei into the muscle cells and increase their capacity of protein synthesis. Testosterone is able to decrease fat formation, having an oxidative effect. The level of testosterone decreases upon long-lasting low intensity training (Hackney, 2001). This type of training does not result in cross-sectional area growth, but it increases upon high intensity training (Vingren et al., 2010). Aerobic training exerts its cardioprotective effect via decreased testosterone level. Upon high intensity training (over 70%), the testosterone level increases for 30–40 min (West and Phillips, 2010). The "Abadjiev method" is a Bulgarian method of powerlifting, which exerts the latter observation, and between 30 min sessions of

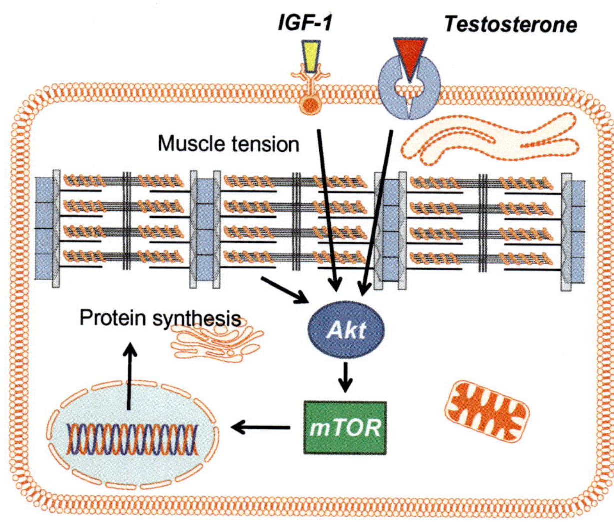

Fig. 4.6 Effect of IGF-1 and testosterone on the sarcomere. Anabolic hormones exert their influence on targeted organs via receptors. In skeletal muscle, IGF-1 and testosterone activate cellular pathways that induce protein synthesis.

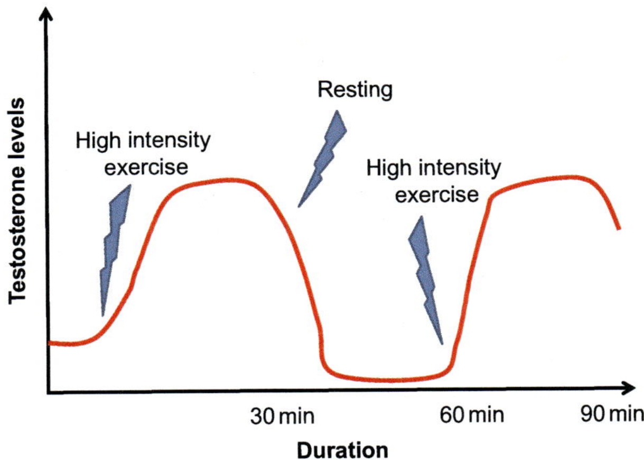

Fig. 4.7 Level of testosterone and duration of training. High intensity training stimulates testosterone production for 30–40 min, then declines. During this decline a resting period promotes muscle growth. This method of training is developed by a Bulgarian trainer, Abadjiev, who had a successful career in powerlifting.

training there is a 30 min resting period, ensuring a high testosterone level during the high intensity training. This method promotes better adaptation to physical training (Fig. 4.7).

Apprehension of the main mechanisms of muscle growth now makes it easier to understand how intensity and resistance training in bodybuilding (e.g., maximal resistance (RM) 60%–80%, 6×8) leads to muscle hypertrophy. Power training promotes special adaptations, and optimal muscle tension is a key factor in trainings targeting muscle growth. The next section discusses the methodology of power training.

4.6 METHODOLOGY OF STRENGTH TRAINING

When selecting a type of training, one needs to set goals. This is the most important and the most difficult part from a trainer's point of view. The aims of strength training can be various, thus we focus on maximal strength, explosive strength, strength endurance, and methods that target cross-sectional area growth and neuromuscular coordination.

4.6.1 Maximal Strength Training

The value of maximal strength is important in many sports since its value can influence both explosive strength and strength endurance. Maximal strength is determined by age, with a peak at the age of 20–35 in males, and 18–30 in females. There are two ways to develop maximal strength: high muscle tension exercises or high speed exercises. High muscle tension exercises stimulate muscle growth if high numbers of serials and repetitions are set (Wernbom et al., 2007). If only 1–2 repetitions are set, improved maximal strength is the result of motor unit recruitment and synchronization. This is important in sports where muscle growth is avoidable, such as high jump or martial arts. In these cases, high intensity exercises (95%–100%) are beneficial. A similar effect can be obtained by maximal speed exercises, where there is low resistance, and the goal is to recruit fast muscle fibers quickly (Hakkinen, 1994). These types of exercises increases maximal strength since more muscle fibers are recruited, and fast fibers are also recruited before all the slow fibers are activated (see details at the beginning of this chapter). Maximal strength can be developed more effectively with heavy weights (80%–100% of individual maximum) and 1–5 repetitions; however, individual differences show variations. One of the most popular methods is the combination of the two methods: high volume moderate resistance exercises (60%–70% of individual maximum, 6–8 repetitions, 6–8 serials) in the first phase, followed by high intensity exercises with low repetition numbers (90%–100%, 1–3/3–5). If there is more than one goal during a single training—for instance, improving maximal strength and also endurance—in this case exercises targeting maximal power development need to be scheduled first.

Some proved method are listed below for maximal strength development; however, there are several other training methods.

Pyramid Method

In this method, resistance needs to be increased continually from 60% to maximal intensity and then reduced (e.g., 60% 1×6, 70% 1×6, 80% 1×5, 90% 1×3, 100% 1×1, 90% 1×3, 80% 1×5, 70% 1×6, and 60% 1×6). There is huge muscle tension during this training, which is essential for muscle hypertrophy (NO and MGF activation), and it may contain maximal speed exercises also (e.g., at 60%–70% resistance), which promote more effective activation of fast motor units. Because of the high intensity during exercises, anabolic hormones promote adaptation processes. In this training, mainly concentric contractions occur, and between sets the athlete needs to rest for 1–3 min. Depending on individual characteristics, age, level of fitness, etc.

Isometric Strength Training

As indicated by the name, there is no joint movement during this type of exercise; strength exertion occurs at an angle. Maximal isometric strength is associated with maximal dynamic strength, thus it is useful in certain sports. The level of tension needs to be close to maximal (90%–100%); if this cannot be measured, one needs to try 100% for 8–12 s 6×8. Tension activates satellite cells and they provide new nuclei via division to muscle fibers, resulting in stimulated protein synthesis and consequential muscle growth. Isometric training has a significant effect on muscle growth in a given joint angle.

Isokinetic Training

Isokinetic training is carried out in a constant speed, with the use of machines. Muscle growth results in response to tension, which need to be close to maximal during exercising. Because there is no free weight in use, it is a perfect method to avoid injury for the elderly. This method is popular among swimmers, kayak and canoe athletes, and rowing athletes, since resistance during these exercises feels similar to that in the water. In the case of lower resistance and higher repetitions, it improves strength endurance.

Bodybuilding Method

This method is used to increase the cross-sectional area. Because there is a close correlation between the cross-sectional area and strength, this method is proper for strength development. The basic concept is to exercise with submaximal resistance with several repetitions and serials (e.g., 70%, 6×6). This leads to satellite cell activation via MGF and NO, and satellite cells provide new nuclei via division to muscle fibers, resulting in stimulated protein synthesis and consequential muscle growth. This type of exercise can be scheduled at the end of the training since there is no need to execute these exercises with high speed or use maximal resistance.

Eccentric Training for Maximal Strength

It has been noted that extreme loading is very effective and it has been used regardless of its high risk of injuries. When the external force on the muscle is greater than the force that the muscle can generate, in eccentric contraction, the muscle is forced to lengthen. Indeed, the highest rate of microscopic injuries occurs in eccentric contractions because of the high external load, thus this type of contraction stimulates the MGF and NO systems the most. NO-dependent follistatin production is an important part of regeneration, which results in contractile protein synthesis and consequential thickening of the sarcomeres.

Eccentric exercises allow joint angle alterations. Lowering the weight back down involves an eccentric contraction, after lifting the weight with the help of training partners; 3–6 repetition and serials are very effective. Because of high resistance, fast fibers are activated almost 100%. This type of exercise is only advisable for trained athletes and after sufficient resting and warming up.

Intermediate Submaximal and Maximal Resistance Training

The effect of this type of training is produced by the mechanisms activated by high resistance. These eccentric and isometric types of method are able to increase not only the amount of contractile fibers but the amount of connective tissue as well. Tendons connect muscle to bones, and transmit forces generated by sarcomeres. The lifetime of collagen and elastin is long (over 200 days), thus synthesis of connective tissue is slower compared to contractile fibers during training targeting muscle hypertrophy. The consequence of the latter is increased risk of injuries, mainly at the muscular-tendon junctions. To avoid injury, one needs to give extra attention to strengthening muscular-tendon junctions by selecting training methods that affect both connective tissue and contractile fibers. The eccentric and isometric types of method are a good selection to develop both components.

4.6.2 Explosive Strength

Explosive strength and reactive or plyometric strength have similar physiological and mechanical characteristics; therefore we discuss them together. Further relevant literature is listed in the bibliography.

During these exercises the goal is to activate and synchronize as many motor units as possible. Unlike maximal strength training where the goal is to induce muscle growth, speed training is aiming at reaching the maximal speed regardless of resistance. Exercises that create special states where potential shortening speed of the muscle is higher, such as pretension, quick-release contractions, and reactive movements, are useful for speed training. This type of movement recruits the connective tissue of the muscles, resulting in higher speed. It is important to note that speed training exercises require a rested condition, and fatigue needs to be avoided. The physiological reason behind that is that fast fibers are fatigable, and so it is impossible to activate and synchronize as many motor units as possible when fatigue occurs. It has been demonstrated that during bench pressing, explosive strength develops to a greater extent with a pop out at the end of a movement compared to just slowly pressing the weight all the way up. During explosive strength training, one needs to focus to execute exercises with a maximal speed regardless of resistance. Thus, speed strength exercises need to be scheduled at the beginning of a training class right after a warm-up.

Dynamic Variable Resistance Method

The point is to carry out the exercises at 30%–80% of maximal strength and with a low repetition number and maximal speed. The reason behind the low repetition number is the fast fibers fatigable nature. Getting up from knee angle at 90° with maximal speed and 60% weight, 5×3 repetitions fosters better innervation and because of the low repetition number it does not promote cross-sectional area growth compared to bodybuilding training. Resistance can be altered in a wide range, but the focus is to strike for maximal speed and avoid fatigue.

Quick-Release Contractions

This method combines isometric and concentric contractions. At the beginning of exercises there is a 3–5 s isometric contraction to increase muscle tension; at a given joint angle contractile elements contract and lengthen elastic elements, thus during contraction the muscle contracts at a high speed. Several motor units are recruited in this way, thus this type of exercise improves neuromuscular coordination. Fast fibers are fatigable, thus this type of exercise prefers low repetitions and enough resting between sets. To carry out these exercises one needs a training partner or a special machine, since isometric contractions need to be executed with maximal intensity or close to it. Thus, it is recommended for professional athletes.

Reactive Training

These exercises start with muscle lengthening by eccentric contraction preceding concentric contraction. The muscle is able to exert partially the mechanical energy accumulated by eccentric contraction, resulting in faster muscle shortening. Exercises such as drop jump, bar jumping, throwing up a ball and catching it behind the head and with the lengthened muscles throwing the ball straight, etc. Athletes should avoid fatigue during these type of exercises because the huge muscle tension and velocity carry a higher risk of injury.

Aiding or Resistance Tools/Competition Situations

To develop explosive strength, tools can be used such as throwing with aiding or resistant tools, throwing a medicine ball, running downhill and uphill, hauling, swimming with a kickboard, swimming against the current, etc. These movements promote the improvement of speed strength since they are speed oriented and resemble the actual movements of a given sport.

Electro Stimulation

This complementary method could also fit the next part (endurance training); however, this method used to be employed in trainings aiming to recruit fast fibers. Muscle stimulation with high frequency, 50–100 Hz through surface electrodes promotes transformation of intermediate fibers to fast fibers. Experiments involving cross-innervation, where slow fiber innervation was replaced by nerves innervating fast fibers via surgical procedures, showed that slow fibers exhibited properties similar to fast fibers following the procedure. Electro muscle stimulation show similar results (Pette et al., 1973). This method is used as a complementary method; however, it targets fast fibers, and intermediate fibers can be transformed to fast fibers. At the case of low frequency (e.g., 5 Hz), intermediate fibers transform to slow fibers, whereas at high frequency (50–100 Hz), they transform to fast fibers.

4.6.3 Strength Endurance Training

Strength endurance involves long-lasting exertion with moderate resistance. It is important in sports played with balls, martial arts, middle- and long-distance sports such as running, cycling, kayaking, and swimming. When designing strength endurance training, one needs to take into account the characteristics of a given sport. For instance, a tennis player who needs the same strength through a game even in the fifth hour and who must hit the ball to reach 200 km/h needs different endurance than an 800 m swimmer or a wrestler, or a kayak athlete competing over 1000 m.

Low Resistance Training

This type of training is the most popular to develop strength endurance (30%–50% resistance, 20–50 repetitions, and 8–20 serials). These exercises need to go on until fatigue develops, unlike during speed strength training. The characteristics of a given sport need to be considered when designing strength endurance training and selecting optimal resistance, and setting repetitions and serials. During these exercises, manly slow fibers are recruited; however, it may require fast fibers as well. Lactic acid production can be significant during this type of training, which may influence resistance against lactic acid and elimination. This topic will be further discussed in Chapter 5.

Circuit Training

This type of training is also popular to develop strength endurance. Included exercises target different muscle groups. Exercises employing own body weight are frequently used and they are safe. This makes circuit training very popular among young athletes. These exercises need to be carried out at low intensity (30%–50%) with high repetitions. There is a significant amount of lactic acid accumulation during this type of training.

4.6.4 Training for Cross-Sectional Area Growth

There is a strong correlation between strength and cross-sectional area, and sometimes it is been interchanged; however, muscle strength can be improved by leaving muscle size unchanged. Increasing the size of a muscle or preventing its deterioration (e.g., in the elderly) can be accomplished by regulating the degree of muscle tension and duration. The more significant the tension is and the longer the duration is, the more effective the muscle hypertrophy training is. Movements need to be executed slowly, since that increases tension. Regardless of the slow velocity, fast fibers in the cross-sectional area increase, mainly because of the thinner Z-line at the border of the sarcomeres, which can be injured by heavy tension. These microscopic injuries serve as adaptation injuries, which are partly prerequisites for the following elevated protein synthesis resulting in elevated resistance of Z-lines to tension. Thus, the same tension next time does not cause injury and there is consequential cross-sectional area growth. Actin and myosin fiber synthesis also occurs in addition to Z-line strengthening.

Temporary muscle growth may occur following training, causing high muscle tension, which is a result of the alterations of liquid volume. This may look like muscle growth, but it is not associated with increased strength compared to real cross-sectional area growth, which is the result of increased protein synthesis. This type of exercise can go on until fatigue develops, or can be carried out during fatigue since cross-sectional area growth is determined by the degree of muscle tension and duration. Up to 16 years of age, training for cross-sectional area growth is not recommended; in any case it would not be effective, since hormonal levels at that age do not support significant muscle growth. Over the age of 60, although increasing muscle size is not impossible, it requires more effort compared to at the age of 30, for example. In this case, hormonal changes play a role as well.

To induce muscle cross-sectional area growth, the most effective methods are bodybuilder training and the isometric method; however, individual differences are possible.

4.6.5 Neuromuscular Coordination Training

Improving neuromuscular coordination is an important goal in several sports (e.g., high jump, volleyball, martial arts), and this type of training is the most challenging for trainers. This type of training requires rested, motivated athletes with excellent concentration since maximal speed is required during exercise. It needs to be carried out with maximal intensity and low numbers of repetitions and serials, thereby activating fast motor units with high activation thresholds. Intermuscular coordination is also very important, thus agonist muscles are supporting each other, and antagonists are in a relaxed state. Preactivated muscle may influence contraction velocity, and adaptation following exercise carried out with maximal velocity results in improved preactivation.

4.6.6 Occlusion Training

Dr. Sato's observation shows that muscles react differently to loading when blood flow is restricted. This method is called occlusion or blood flow restriction (BFR), and during exercise vasculature is compressed proximal to the exercising muscle, which produces local ischemia in the limb (Fig. 4.8). It can be achieved using a simple rubber band or a tourniquet to control blood flow. The extent of occlusion is an important point, and even 20% resistance can result in cross-sectional muscle growth because of the metabolic changes in the muscle. It is associated with increased GH production, and lower degree of protein degradation, and gene expression with consequential protein synthesis. Occlusion training needs medical attendance,

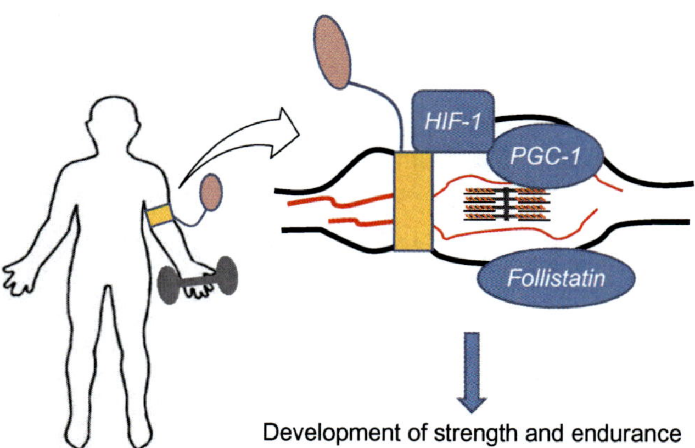

Fig. 4.8 Occlusion training. During occlusion training, a tourniquet is placed on the muscle to reduce blood flow. This induces local ischemia resulting in metabolic changes, and cross-sectional muscle growth following even a lower resistance training. It is interesting to note that strength endurance also improves in addition to muscle growth. This method activates hypoxia inducible factor 1 (HIF-1) and peroxisome proliferator-activated receptor gamma coactivator 1-alpha (PGC-1 alpha); the latter is a transcriptional cofactor responsible for mitochondrial biogenesis.

and it can be very useful in rehabilitation, cross-sectional area growth, and strength endurance development (Item et al., 2013).

4.6.7 Unstable Surface Training (Proprioception Training)

Proprioception comes from a Latin word meaning unconscious perception of movement. It allows the body to control its position for optimal locomotion. It is carried out by internal sensors such as the muscle spindle stretch receptor and Golgi tendon organ. The vestibular system in the brain is a key component in proprioception and also in maintaining static, mixed, or dynamic balance. Proprioception training improves balancing, movement sensing, and, naturally, proprioception.

Proprioception is present in every muscle movement, therefore proprioception training may be misleading. Unstable surface training is a more expressive name for this type of method. Proprioception is extremely important in motor learning, smooth motor learning, and preventing injury (Verhagen et al., 2004). The latter observation made unstable surface training very popular, where movements are carried out in a position that needs constant balancing. Exercises on a wobble board improve ankle and knee functional stability and prevent injuries (Cloak et al., 2013; Sparkes and Behm, 2010). A balance board or unstable straps push-up exercises used for shoulder joint training give different EMG signals compared to push-ups executed on a stable surface. On a stable surface, an EMG indicates lower muscle activity (Snarr and Esco, 2013). Unstable surface training is recommended for injury prevention, rehabilitation, and improving physical performance.

4.7 "R" EXTRA

Molecular mechanisms of cross-sectional area growth involve the mitogen-activated protein kinase (MAPK) pathway. Mechanical stress and tension activate cellular signal molecules and consequential protein synthesis. In this pathway, MAPK is in a key position integrating signals from different sources (e.g., hormonal or NO-activated pathways). The MAPK pathway includes c-Jun NH_2-terminal kinase (JNK), which is considered a stress-activated kinase, and also ERK kinases specialized for external signals, and p38 (38 kDa protein). It has been demonstrated that activation and phosphorylation of JNK, ERK, and MAPK are lowest during maximal tension in concentric contractions, higher in isometric contractions, and the highest in eccentric contractions (Martineau and Gardiner, 2001).

The higher tension shows higher MAPK activity compared to lower tension for longer duration, indicating that the degree of tension is a key factor in muscle hypertrophy, and different types of muscle activity stimulate MAPKs and consequential gene expression differently.

Another important factor in muscle hypertrophy is mTOR protein (mammalian target of rapamycin), which integrates signals from growth factors (e.g., GH, IGF-1), amino acids, and metabolic processes, and initiates signaling that results in increased protein synthesis. It influences protein synthesis via the phosphorylation of Akt/PKB proteins. It is interesting that aging-related muscle deterioration is associated with decreased mTOR activity. Thus, activation of the mTOR pathway stimulates muscle growth. A recently discovered subunit (PGC-1α-4) of PGC-1α is able to activate the mTOR pathway, thereby increasing cross-sectional area growth.

Skeletal muscle development requires myogenic differentiation of mesodermal cells. Characteristics of a muscle cell develop via specific gene activation that plays a role in multinuclear muscle fiber development. Transcription factors such as MyoD, Myf5, and myogenin (called muscle regulatory factors, or MRF) are important factors in muscle development (Perry and Rudnick, 2000). Myoblast differentiation and division are regulated by stimulatory and inhibitory proteins. Myostatin is an inhibitory protein belonging to the transforming growth factor-β (TGF-β) protein family. Myostatin gene mutation manifests in greater muscle mass. This observation initiated the research on myostatin inhibitors (e.g., follistatin) and antagonists. Myostatin exerts its influence on muscle differentiation through MyoD inhibition (via Smad 3 protein) (Langley et al., 2002). Follistatin activates nNOS, which results in an increased amount of NO and consequential stimulation of satellite cell division, which may play a role in cross-sectional area growth. Muscle growth can be accomplished without satellite cell activation, for instance, through increased protein synthesis of the existing nuclei in muscle fibers (Fig. 4.9). Thus, it does not require an increased number of nuclei.

There is a debate about hyperplasia, cell proliferation, and new muscle cell formation. Muscle and nervous cells are not able to divide, thus according to recent knowledge, biogenesis of new cells in these tissues is not possible. Thus, adaptation in muscle manifests in hypertrophy, rather than increased number of muscle fibers (Fig. 4.10). However, several observations have shown formations of new cells, and following significant muscle tension resulted in hyperplasia in addition to hypertrophy. Hyperplasia was observed in some human studies following anabolic steroid intake. According to the criticals of hyperplasia, there is no such method to count the exact number of all cells (e.g., biopsies taken from

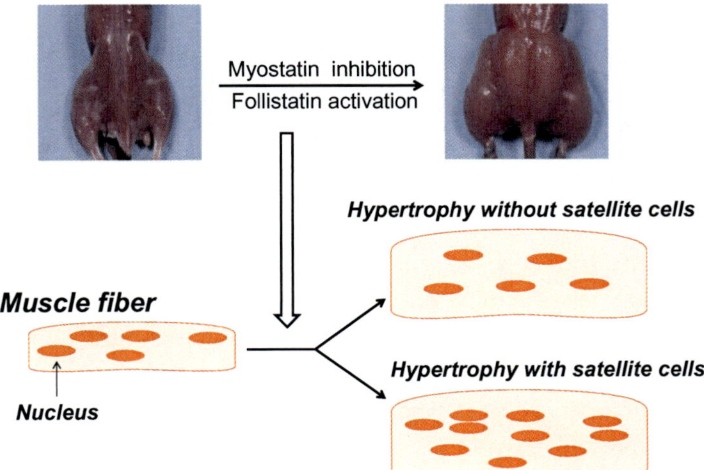

Fig. 4.9 Cross-sectional area growth and satellite cell activation. Myostatin has an important regulatory function in skeletal muscle development, as it inhibits myoblast division. Myostatin gene mutation results in spectacular muscle growth and enormous strength. Follistatin antagonizes the effect of myostatin, resulting in increased muscle growth. Thus, stimulation of follistatin via gene manipulation or by pharmacological means results in muscle growth. This would be beneficial in the elderly to prevent muscle loss. Follistatin administration, however, is considered doping in professional sports.

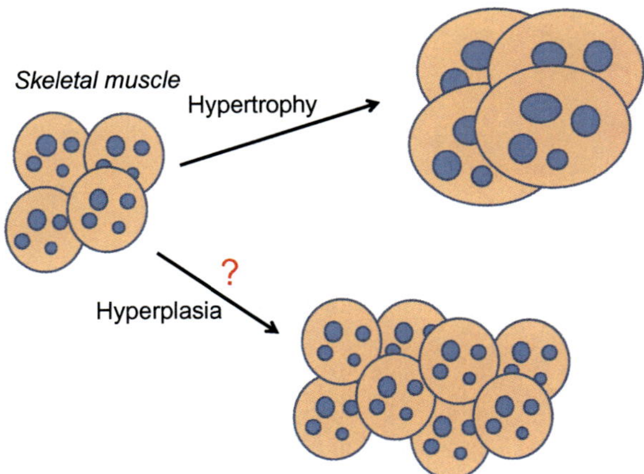

Fig. 4.10 Hypertrophy and hyperplasia. In hypertrophy, cross-sectional area growth is an adaptive process in skeletal muscle following loading. In contrast, there is a debate on hyperplasia (increased number of muscle fibers), regarding whether it is an adaptive response or a degenerative process.

different places may result in false counts). Thus, differences in cell counts may be a result of an inaccurate biopsy rather than cell division. Hyperplasia is considered as either a degenerative process or a physiological and/or regenerating process.

4.8 SUMMARY

Strength is determined by several factors such as age, sex, arrangement of muscle fibers, cross-sectional area of muscles, number of recruited motor units and their synchronization, and type of muscle contraction. Types of strength include maximal strength, explosive strength, and strength endurance, which can be improved by different types of training. In muscle growth, satellite cells play a role, and are activated by mechanical stress, tension, and injury. Important factors in hypertrophy include nitrogen oxide, follistatin, GH, IGF-1, and testosterone. Activation of motor units depends on their activation thresholds; however, recruitment of fast fibers is possible during quick movements. In muscle growth training, the degree of tension and its duration are key factors that induce adaptation, while in neuromuscular coordination training, recruitment of motor units is the most important factor (Fig. 4.11).

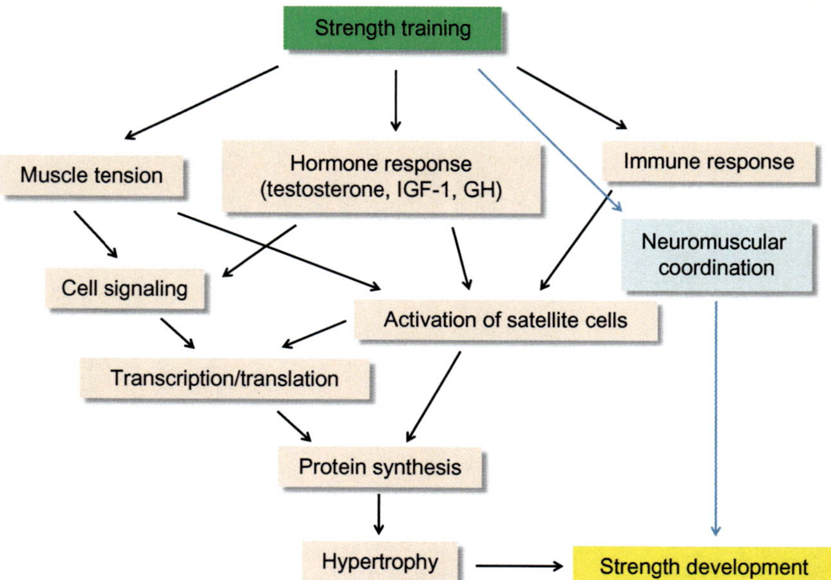

Fig. 4.11 Strength development sequence. Improved strength is a result of increased cross-sectional area (size) of muscles and improved neuromuscular coordination.

TEST QUESTIONS

1. What factors determine the force generated by the muscle?
2. What is the mechanism of cross-sectional area growth of muscles?
3. How are motor units recruited during different loading?
4. Explain the correlation between different types of muscle contractions and the force generated by the muscle.
5. What are the characteristics of the maximal strength training method?
6. What are the differences between the maximal strength and speed strength training methods?
7. What are the most important factors in strength endurance development?
8. What is the difference between neuromuscular coordination training and training for cross-sectional area growth?
9. Why and how does age affect adaptation to strength training?

BIBLIOGRAPHY

Bruusgaard, J.C., Johansen, I.B., Egner, I.M., Rana, Z.A., Gundersen, K., 2010. Myonuclei acquired by overload exercise precede hypertrophy and are not lost on detraining. Proc. Natl. Acad. Sci. U. S. A. 107, 15111–15116.

Cloak, R., Nevill, A., Day, S., Wyon, M., 2013. Six-week combined vibration and wobble board training on balance and stability in footballers with functional ankle instability. Clin. J. Sport Med. 23, 384–391.

Collier, S.R., Casey, D.P., Kanaley, J.A., 2005. Growth hormone responses to varying doses of oral arginine. Growth Horm. IGF Res. 15, 136–139.

D'Antona, G., Lanfranconi, F., Pellegrino, M.A., Brocca, L., Adami, R., Rossi, R., Moro, G., Miotti, D., Canepari, M., Bottinelli, R., 2006. Skeletal muscle hypertrophy and structure and function of skeletal muscle fibres in male body builders. J. Physiol. 570, 611–627.

Duchateau, J., Semmler, J.G., Enoka, R.M., 2006. Training adaptations in the behavior of human motor units. J. Appl. Physiol. 101, 1766–1775.

Fling, B.W., Christie, A., Kamen, G., 2009. Motor unit synchronization in FDI and biceps brachii muscles of strength-trained males. J. Electromyogr. Kinesiol. 19, 800–809.

Frystyk, J., 2010. Exercise and the growth hormone-insulin-like growth factor axis. Med. Sci. Sports Exerc. 42, 58–66.

Hackney, A.C., 2001. Endurance exercise training and reproductive endocrine dysfunction in men: alterations in the hypothalamic-pituitary-testicular axis. Curr. Pharm. Des. 7, 261–273.

Hakkinen, K., 1994. Neuromuscular fatigue in males and females during strenuous heavy resistance loading. Electromyogr. Clin. Neurophysiol. 34, 205–214.

Herrmann, U., Flanders, M., 1998. Directional tuning of single motor units. J. Neurosci. 18, 8402–8416.

Item, F., Nocito, A., Thony, S., Bachler, T., Boutellier, U., Wenger, R.H., Toigo, M., 2013. Combined whole-body vibration, resistance exercise, and sustained vascular occlusion increases PGC-1alpha and VEGF mRNA abundances. Eur. J. Appl. Physiol. 113, 1081–1090.

Kostka, T., Patricot, M.C., Mathian, B., Lacour, J.R., Bonnefoy, M., 2003. Anabolic and catabolic hormonal responses to experimental two-set low-volume resistance exercise in sedentary and active elderly people. Aging Clin. Exp. Res. 15, 123–130.

Langley, B., Thomas, M., Bishop, A., Sharma, M., Gilmour, S., Kambadur, R., 2002. Myostatin inhibits myoblast differentiation by down-regulating MyoD expression. J. Biol. Chem. 277, 49831–49840.

Lu, H., Huang, D., Saederup, N., Charo, I.F., Ransohoff, R.M., Zhou, L., 2011. Macrophages recruited via CCR2 produce insulin-like growth factor-1 to repair acute skeletal muscle injury. FASEB J. 25, 358–369.

Maejima, H., Murase, A., Sunahori, H., Kanetada, Y., Otani, T., Yoshimura, O., Tobimatsu, Y., 2007. Neural adjustment in the activation of the lower leg muscles through daily physical exercises in community-based elderly persons. Tohoku J. Exp. Med. 211, 141–149.

Martineau, L.C., Gardiner, P.F., 2001. Insight into skeletal muscle mechanotransduction: MAPK activation is quantitatively related to tension. J. Appl. Physiol. 91, 693–702.

McDonagh, J.C., Binder, M.D., Reinking, R.M., Stuart, D.G., 1980. A commentary on muscle unit properties in cat hindlimb muscles. J. Morphol. 166, 217–230.

McDonald, K.S., Wolff, M.R., Moss, R.L., 1997. Sarcomere length dependence of the rate of tension redevelopment and submaximal tension in rat and rabbit skinned skeletal muscle fibres. J. Physiol. 501 (Pt 3), 607–621.

Nishida, Y., Matsubara, T., Tobina, T., Shindo, M., Tokuyama, K., Tanaka, K., Tanaka, H., 2010. Effect of low-intensity aerobic exercise on insulin-like growth factor-I and insulin-like growth factor-binding proteins in healthy men. Int. J. Endocrinol. 2010.

Ogasawara, R., Kobayashi, K., Tsutaki, A., Lee, K., Abe, T., Fujita, S., Nakazato, K., Ishii, N., 2013. mTOR signaling response to resistance exercise is altered by chronic resistance training and detraining in skeletal muscle. J. Appl. Physiol. 114, 934–940.

Perry, R.L., Rudnick, M.A., 2000. Molecular mechanisms regulating myogenic determination and differentiation. Front. Biosci. 5, D750–767.

Pette, D., Smith, M.E., Staudte, H.W., Vrbova, G., 1973. Effects of long-term electrical stimulation on some contractile and metabolic characteristics of fast rabbit muscles. Pflugers Archiv: Eur. J. Physiol. 338, 257–272.

Snarr, R.L., Esco, M.R., 2013. Electromyographic comparison of traditional and suspension push-ups. J. Human Kinetics 39, 75–83.

Sparkes, R., Behm, D.G., 2010. Training adaptations associated with an 8-week instability resistance training program with recreationally active individuals. J. Strength Cond. Res. 24, 1931–1941.

Verhagen, E., van der Beek, A., Twisk, J., Bouter, L., Bahr, R., van Mechelen, W., 2004. The effect of a proprioceptive balance board training program for the prevention of ankle sprains: a prospective controlled trial. Am. J. Sports Med. 32, 1385–1393.

Vingren, J.L., Kraemer, W.J., Ratamess, N.A., Anderson, J.M., Volek, J.S., Maresh, C.M., 2010. Testosterone physiology in resistance exercise and training: the up-stream regulatory elements. Sports Med. 40, 1037–1053.

Wernbom, M., Augustsson, J., Thomee, R., 2007. The influence of frequency, intensity, volume and mode of strength training on whole muscle cross-sectional area in humans. Sports Med. 37, 225–264.

West, D.W., Phillips, S.M., 2010. Anabolic processes in human skeletal muscle: restoring the identities of growth hormone and testosterone. Phys. Sportsmed. 38, 97–104.

Zajac, F.E., Faden, J.S., 1985. Relationship among recruitment order, axonal conduction velocity, and muscle-unit properties of type-identified motor units in cat plantaris muscle. J. Neurophysiol. 53, 1303–1322.

CHAPTER 5

Fundamentals of Endurance Training

The endurance of the human body is remarkable—for instance, humans are more enduring in exposure to continuous loading than horses. Yiannis Kouros, an ultramarathon runner, completed 303 km a day, while Haile Gebrselassie completed a 42,195 m marathon distance at 20 km/h. In the Sydney Olympics, Gebrselassie ran the 10,000 m in 27:18.20; behind him was Paul Tergat 27:18.29, and Assefa Mezgebu at 27:19.75. Thus, the difference between first and third place was only 0.04%, indicating that there is only a slight difference between the best athletes, which raises the necessity of effective training.

Endurance can be trained and developed; however, genetic factors are also important. The dominance of East African athletes in aerobic sport categories shows the importance of genetic factors in endurance, which will be discussed in Chapter 11.

The adaptation of human body to endurance training differs from the reaction to strength training, since endurance training induces metabolic changes and cardiovascular responses. Endurance usually refers to aerobic endurance; however, anaerobic endurance is just as important in sports.

Oxygen consumption (VO_2) during training can be described by the Fick equation:

$$VO_2 = \text{cardiac output}\,(Q) \times A - VO_2\,\text{difference}$$

Thus, maximal oxygen consumption is equal to cardiac output multiplied by arterial-venous concentration difference of oxygen. This equation indicates that cardiac output is an important limiting factor of maximal oxygen consumption. Other factors influence maximal oxygen uptake such as number of red blood cells (RBC), muscle vascularization, mitochondrial supply of the muscle, and activity of enzymes.

Since VO_2 is determined largely by body size, therefore relative maximal oxygen (VO_2max) consumption is used as an objective indicator; this is maximal oxygen consumption divided by body weight. VO_2max is an

The Physiology of Physical Training
https://doi.org/10.1016/B978-0-12-815137-2.00005-X
 81

important indicator not only in sports but in disease prevention, which will be discussed in detail in Chapter 9.

A.V. Hill and his colleague Lupton showed in the 1920s that oxygen consumption determines physical performance, and although Hill's theory is debated, no data have proved otherwise (Hill and Lupton, 1923). Endurance is determined by other factors in addition to cardiovascular factors, such as musculature, the central nervous system, psychological factors, and metabolic factors. Energy production is an important limiting factor, which is determined by metabolic processes (aerobic, anaerobic lactic, and anaerobic alactic). Continuous contractions for 10 min in the quadriceps femoris consume 5 millimoles (approx. 0.9 g) glucose; 22%–26% is used for actual work because of the skeletal muscle efficiency, while the rest is used in heat generation. Efficiency of skeletal muscle is better than F1 motors (Richter et al., 1988). During long-lasting training significant heat production challenges the body, and effective heat elimination during contraction is essential for optimal performance. The human body obtained effective heat elimination after losing its hair coat, which may be the evolutionary cause. Long-distance runners from Kenya have better heat elimination efficiency than their European counterparts with similar VO_2max.

Endurance is also a complex ability since successfulness is also determined by the efficiency of locomotion, which shows individual differences. The next paragraphs discuss the physiological determining factors, followed by endurance training methods. The complex factors determining endurance are shown in Fig. 5.1.

5.1 THE HEART AS A MAIN FACTOR IN ENDURANCE PERFORMANCE

The average adult human heart weighs 250–350 g and is as big as a fist. The wall of the heart is made up of three layers: epicardium; myocardium or cardiac muscle; and endocardium, which is in direct contact with blood. The heart has four chambers: two upper atria and two lower ventricles. During contraction of the ventricles (systole), blood flows into the vessels accompanied by increasing blood pressure. From the left ventricle, blood flows to the body via the aorta, and from the right ventricle to the lungs. Relaxation of the ventricles (diastole) is accompanied by decreasing blood pressure in the vessels and allows blood to flow from the two atria. The blood flows through the tricuspid valve from the right atrium to the right ventricle, and via the mitral valve from the left atrium to the left ventricle

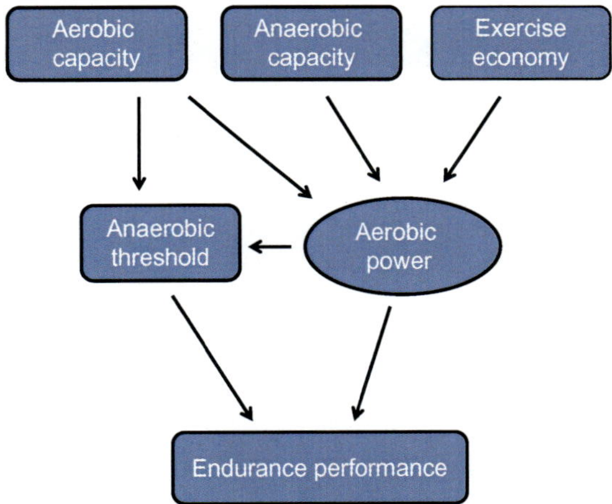

Fig. 5.1 Endurance performance. Endurance performance includes aerobic and anaerobic metabolic factors and efficiency of locomotion.

to prevent backflow. Oxygenated blood from the lungs flows through the left atrium to the left ventricle, then through the semilunar aortic valve to the aorta. The walls of the ventricles are thicker since they pump the blood against high pressure. The left ventricle pumps the oxygenized blood toward the organs of the body through the aorta and vessels with million capillaries. The deoxygenized CO_2-rich blood flows from the organs back to the right atrium (Fig. 5.2). The heart and its functional capacity is the main determinant of aerobic endurance. The size of the left ventricle and

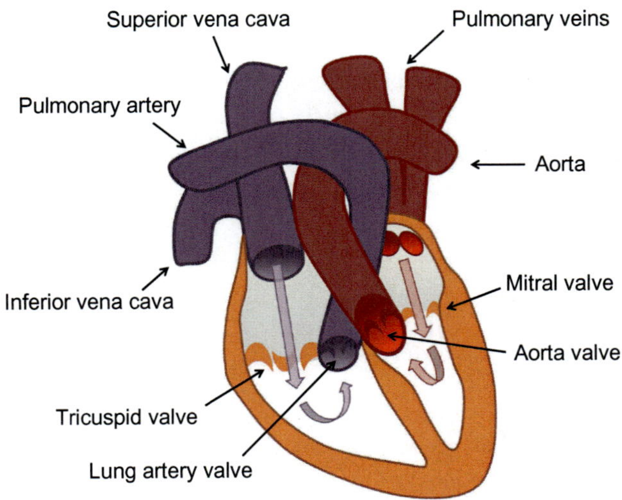

Fig. 5.2 The heart. Simplified anatomy of the human heart.

the volume it can pump out per minute can be increased by moderate intensity training, while high intensity training with submaximal or maximal heart rate (e.g., during strength training) does not result in increased cardiac output (heart rate × stroke volume). The difference between resting heart rate and maximal heart rate shows the physical fitness of the body. Maximal heart rate is calculated as 220 minus age. With excellent physical fitness, the resting heart rate is lower (below 50), which is called bradycardia.

The heart muscle is homogeneous; unlike skeletal muscle, it is not comprised of different types of motor units. Thus, when a cardiac muscle cell is activated, activation flows through the rest of the cells. The cardiac muscle contracts autonomously even without any outside stimuli. The heart's pacemaker is the sinoatrial node, and these cells determine the resting heart rate. The electrical signal travels through the atrioventricular node and through the heart, causing the cardiac muscle to contract. The signal is delayed slightly (approx. 0.1 s) at the atrioventricular node to allow the ventricle to fill up with blood before contraction of the ventricles. The signal reaches the whole ventricle via Purkinje fibers and then this contracts and pumps out the blood. The electrical signal can be recorded via an electrocardiogram (ECG), which provides information about the heart's well-being. The average resting heart rate of an adult is 68–72 per min, and the maximal heart rate, as noted above, is calculated as 220 minus age. Children have higher values, and maximal heart rate is decreases with age.

An ECG registers electrical signals of the heart (Fig. 5.3), and an exercise ECG provide abundant information about heart functions and shows how the heart responds to exertion. Modern imaging techniques such as ultrasound, echocardiography, and magnetic resonance imaging (MRI) provide extra information in addition to an ECG.

Adaptation to training can be illustrated by heart rate changes. Decreased heart rate of trained individuals provides a greater range for changes during training since maximal heart rate is not affected by fitness; thus, higher difference between resting heart rate and maximal heart rate provide a wider zone for adaptation responses. Trained endurance athletes can have a resting heart rate of under 40 beat/min. Another indicator in addition to heart rate reserve (the difference between resting heart rate and maximal heart rate) is the heart rate recovery to the resting value, which is quicker in trained individuals.

It is interesting that adaptation to altitude results in decreased maximal heart rate. Atmospheric pressure decreases at high altitudes, resulting in impaired oxygenation of the hemoglobin. Above 2000 m, maximal heart rate

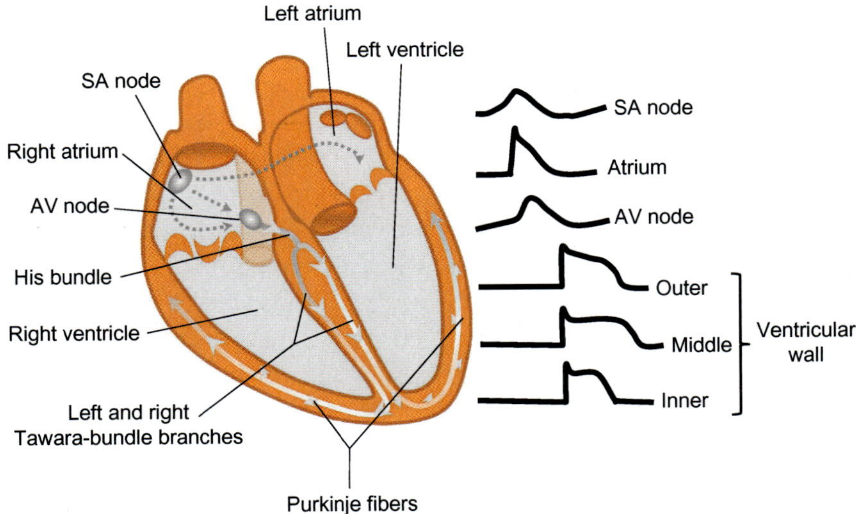

Fig. 5.3 Electrical signals of the heart and contraction. Action potentials from different cardiac cells are shown on the right. The electrical signals are recorded as an electrocardiogram (ECG), which provides information about the heart functions and shows how the heart responds to exertion.

decreases, and is significant above 6000 m. Decreased maximal heart rate provides a longer time for oxygenation in the lungs. If the maximal heart rate does not change, for instance, 200 beat/min. Would not allow enough time for oxygenation in the lung. A high altitude impairs endurance because of the narrowed range for changes (increased resting heart rate and decreased maximal heart rate). High-altitude training will be discussed in detail later.

Adaptation of the heart following training is quick as demonstrated by modern imaging techniques, and it fits the stimuli-specific adaptation theory. For instance, following 90 days of sport-specific training, endurance athletes such as rowing athletes showed different adaptations compared to American football athletes, who received high intensity strength training. The ventricles of the heart enlarged in endurance athletes, while strength training induced wall thickening of the left ventricle. In addition, endurance training resulted in a higher degree of cardiac muscle relaxation and consequential higher capacity to receive a higher volume of venous blood; thus, it induces higher tension, which causes hypertrophy. In contrast, the latter results were not noticed following strength training (Baggish et al., 2008). The experimental results were not surprising; however, the fact that it occurred in such a short time (90 days) was unexpected, and indicated that frequent physical training has huge potential.

Enlargement of the heart occurs mainly in response to pathophysiological disturbances, thus enlargement of the heart does not always mean better physiological function. However, the underlying causes differ from those of physiological training. Although both are the results of adaptation processes, heart functions are impaired under pathological circumstances, whereas a healthy heart shows functional improvement in response to training.

During intensive training, oxygen demand of the heart is also as is oxygen demand of the skeletal muscle. For instance, oxygen demand of the left ventricle increased six-fold, thus frequent training results in abundant vascularization of the left ventricle, improving its oxygen supply (Duncker and Bache, 2008).

Increased oxygen demand by the skeletal muscle during training is fulfilled by the elevated performance of the heart and the hormonal system. As a result of adaptation, parasympathetic effects are augmented, resulting in a decreased resting heart rate. Thus, trained athletes also have a lower heart rate during loading, and the saturation of the ventricles is elevated during diastole, which results in increased stroke volume. Higher volume causes higher tension, resulting in cross-sectional area growth of the wall of the left ventricle. It has eccentric characteristics in endurance athletes, and concentric characteristics in strength athletes (Pluim et al., 2000); however, in both cases it is the result of physiological and not pathophysiological changes. During strength training, tension in the muscles compresses the vasculature, thus the heart works against higher resistance, resulting in enlargement of the cardiac muscle. An analogy is to blow an air-balloon with a tight whole representing a high peripheral resistance. During aerobic training, peripheral resistance is not significant compared to strength training (Fig. 5.4).

The results of adaptation, which are also influenced by genetic factors, are increased stroke volume up to 200 mL, increased cardiac output up to 50 L/min, and 90 mL/kg/min VO_2max (Table 5.1). In addition, blood flow is increased in muscles up to 380 mL/min per 100 g tissue, which is a spectacular increase compared to resting muscle, where blood flow is 20–25 mL/min/100 g tissue. These VO_2max and blood flow data show a significant degree of adaptation to physical training in the heart.

Endurance training is very important in every sport and age because of its beneficial effect on the heart. Aerobic training is also important for power athletes, where loading is high for a short duration, because they also need endurance during the long training sessions, which may bring success. A weight-lifting athlete needs to train for long hours and days to perform well during those few seconds of performance at a competition, and he or she needs endurance. Endurance training also has significant preventive effects, which will be discussed in Chapter 9.

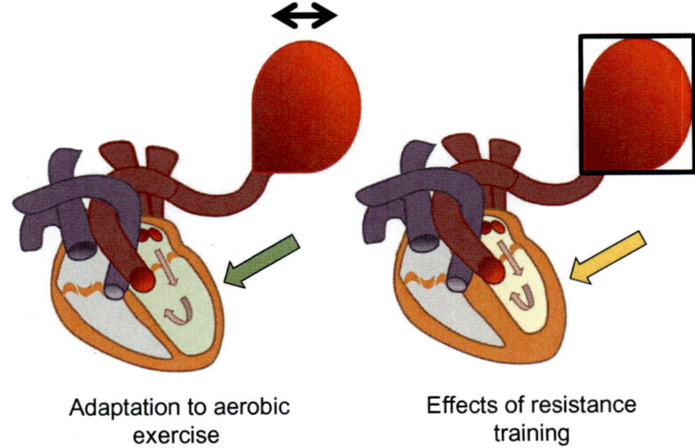

Fig. 5.4 Effects of endurance and strength training on the heart. The heart adapts differently to endurance and strength training. Aerobic adaptation results in enlargement of the left ventricle and decreased resting heart rate. Strength training results in thickening of the left ventricular wall in response to increased peripheral resistance.

Table 5.1 Main indicators of the cardiovascular system

	Rest	Maximal loading		
		Untrained	**Trained**	**Well trained**
Ventilation (L/min)	12–15	120–130	180–200	220–230
Cardiac output (L/min)	5	25	35	40
Heart rate (beat/min)	70	200	200	200
Stroke volume (mL/min)	70	140	180	200
VO_2max (mL/kg/min)	35	50	70	85
Anaerobic threshold ($VO_2max\%$)	–	65%	75%	85%

Sudden cardiac death may occur even in professional athletes, who could be saved by screening. Sudden cardiac death is usually a result of cardiovascular diseases when the athlete is over 35 years old (Durakovic et al., 2002). In younger athletes, it is usually the consequence of arrhythmogenic right ventricle cardiomyopathy, or myocarditis (Lauschke and Maisch, 2009). Since arrhythmogenic right ventricle cardiomyopathy can be screened and prevented, athletes need to be checked before entering professional sports, especially those who have a family history of this.

The heart is the focus during endurance training since maximal oxygen uptake is determined by the cardiac output. Although the heart adapts quickly to changing environments, individual differences, individual maximal values, and adaptation are also determined by genetic factors.

5.2 THE CARDIOVASCULAR SYSTEM AS A DETERMINING FACTOR OF ENDURANCE PERFORMANCE

5.2.1 Blood as Oxygen Carrier and a Determining Factor of Performance

An average adult male has 5 L of blood, which is approximately 8% of total body weight. It contains blood cells, and liquid (plasma) made up of 98% water, which transports oxygen, hormones, and nutrition, and plays a role in immune responses. Blood plasma is about 55% of total blood; the rest is blood cells including RBC, white blood cells, and platelets. Blood supply during training is altered; it is increased in the muscles and the heart, and slightly in the brain, whereas it is decreased in visceral tissues. Therefore the higher oxygen demand upon physical training is fulfilled, which can be a hundred-fold higher in skeletal muscle compared to resting state.

Endurance is determined mainly by RBC (or erythrocytes), which are disc-shaped, flexible cells; their lifetime is 100–120 days, and their concentration is 5 million/μL. RBC are produced by bone marrow, and their differentiation is stimulated by erythropoietin (EPO), a hormone produced by the kidney. RBC do not have nuclei, thus no DNA content, and they are not able to divide. They have a flexible structure, and can therefore flow easily through capillaries; this feature deteriorates with their age, as does their oxygen transporting ability. Their membrane deteriorates with time because of free radical–induced damage, thus they are engulfed by macrophages and degraded by the spleen and liver.

The main component of RBC is hemoglobin, which is reflected by the red color of the blood, and the iron atom embedded in the heme provides anoxygen binding capacity. Upon oxygen binding, it is bright red (oxyhemoglobin), and when deoxygenated (deoxyhemoglobin) it is deep red. Oxygen is always bound to iron-containing molecules in the body. When blood flow velocity increases, or the number of RBC is high due to EPO administration, higher viscosity stimulates endothelial nitrogen oxide synthase (eNOS) resulting in NO production and consequential vascular relaxation (Fig. 5.5). Inhibition of eNOS decreases blood flow in the muscle, especially in vasculature supplying slow-twitch muscle fibers. Thus, NO production due to higher blood flow velocity and viscosity plays a role in better blood supply and better aerobic performance. Frequent physical training increases the availability of NO to smooth muscle cells surrounding the blood vessels.

The percentile of RBC called the hematocrit level is, under normal circumstances, 44%–48% in males and 38%–40% in females. RBC isolated by centrifugation can be stored in a refrigerator and injected back into increase

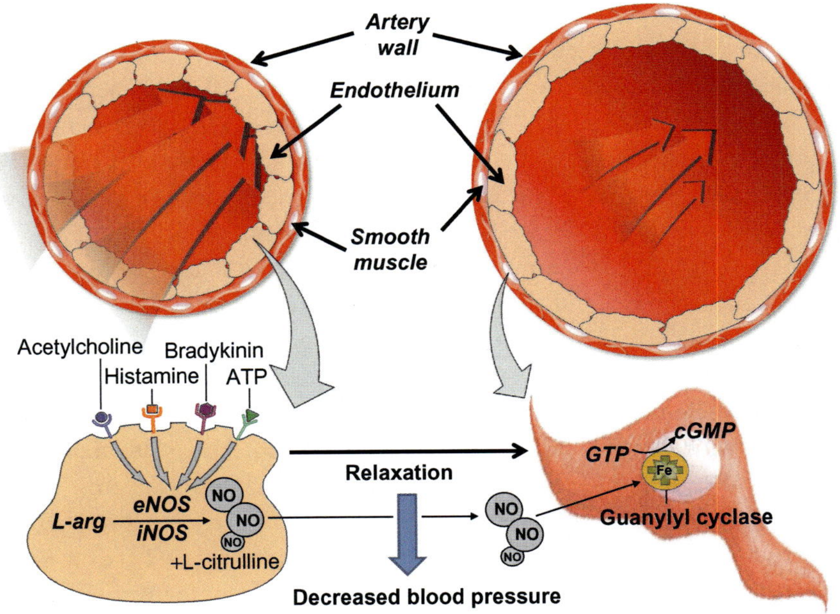

Fig. 5.5 Vascular relaxation. The inner endothelial layer of blood vessels contains endothelial nitrogen oxide synthase (eNOS). When blood flow velocity increases, increased viscosity stimulates NO production in endothelial cells, which consequentially relaxes smooth muscle cells around the blood vessels resulting in increased lumen, thus blood flow velocity and blood pressure decrease.

oxygen transport capacity; this is called blood doping and is illegal in sports. The hematocrit level increases over 60% following EPO administration resulting in increased viscosity, and may result in cardiovascular disturbances, or more serious consequences.

The number of RBC determines oxygen transport efficiency, which is impaired in anemia and iron deficiency, resulting in impaired endurance. In contrast, an increased number of RBC following high–altitude training, or EPO, or blood doping, improves endurance.

Blood pH indicating acidity or alkalinity under normal circumstances in resting state is 7.35–7.45; this is slightly alkaline since 7 is neutral. Following training the blood pH turns to slightly acidic, and can drop below 6.8 in trained athletes. Lactic acid is the main contributor to acidic blood pH, and can be measured by one drop of blood from the fingertip or earlobe. Lactic acid blood concentration in resting state is 1–1.5 mmol/L. During physical training, its peak value is 20–22 mmol/L in trained athletes, whereas in untrained individuals it does not reach 10 mmol/L.

Anaerobic endurance training increases insensibility to lactic acid and trained athletes can tolerate either high or low lactic acid concentration; they also normalize its level more efficiently. Maximal lactic acid level is measured 2–5 min, after loading; during this time lactic acid is transferred to the bloodstream from muscles. Acidic pH has an adverse effect on physiological processes, since biochemical reactions catalyzed by enzymes require an optimal pH. Under certain pH several enzymes are not able to function optimally and impaired metabolic processes may result in fatal consequences. Thus, elimination and neutralization of lactic acid need to be quick and efficient, and indeed it returns to normal level in 30 min. If this were not the case, it would be impossible for a kayak athlete with a lactic acid level of 20 mmol/L to compete in one race followed by another during a competition. Lactic acid is transported by monocarboxylate transporter (MCT) from muscle to blood, and is normalized by two systems. One is lactate dehydrogenase (LDH); this enzyme converts lactic acid to pyruvate under aerobic conditions. During anaerobic glycolysis LDH works in the reverse, converting pyruvate to lactic acid, thus LDH isoforms can catalyze lactic acid–pyruvate conversion both ways.

During intense physical training, a significant amount of lactic acid is produced in skeletal muscle, while during rest lactic acid is converted back to glucose via Cori-circle in the liver (and to some extent also in the muscle). This is the second elimination system, and the Cori-circle comprises steps which are in reverse order compared to anaerobic glycolysis, where glucose is metabolized to lactic acid, then pyruvate, producing 2 ATP. Pyruvate is then converted back to lactic acid by LDH, transferred into the bloodstream, and taken up by the liver, where it is converted back to glucose using 6 ATP (gluconeogenesis).

Blood glucose level is also an important factor in performance since low level (hypoglycemia) induces fatigue, which needs to be prevented (e.g., a tennis player might consume a banana or glucose candy after a game). Following food intake, blood sugar level increases, leading to increased production of insulin. This hormone stimulates glucose uptake by the muscles, thereby decreasing blood sugar level. Fatigue following low intensity training may result in decreased blood sugar level. In contrast, high intensity training (VO_2max over 80%) increases blood sugar level by mobilizing carbohydrates from the liver and the muscle since carbohydrates are the main energy source for ATP synthesis. High intensity training can even cause a hyperglycemic state (Marliss and Vranic, 2002). The underlying cause of the latter is fatigue, which results in decreased glucose

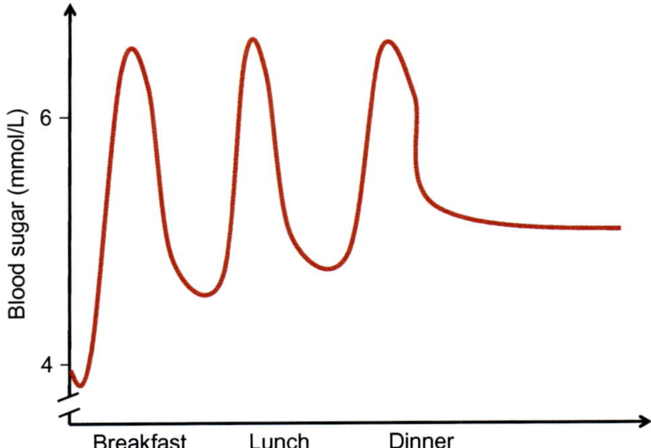

Fig. 5.6 Daily fluctuation of blood sugar level. Blood sugar level is increased upon food intake (breakfast, lunch, dinner), and is also influenced by physical training.

consumption; however, mobilization is running high, thereby causing a high blood glucose level.

Blood sugar level oscillation during a day is dependent on food intake (Fig. 5.6) and physical training. Both high and low blood sugar levels are limiting factors of physical performance. For individuals who suffer from diabetes, physical training is also important to control long-term complications; however, individual differences need to be taken into account during loading. Fasting blood sugar level in individual with type 2 diabetes can reach over 10 mmol/L. A blood sugar level over 30 mmol/L for several hours can result in fatal consequences.

5.2.2 Arterial-Venous Concentration Difference of Oxygen (A-VO$_2$ difference): Efficiency

Arterial-venous concentration difference of oxygen, A-VO$_2$ difference, indicates oxygen difference as milliliter per 100 mL blood. This indicates the amount of oxygen that diffuses into the tissues from the capillaries. Arterial blood is taken from the thigh arteria, while venous blood is taken from the pulmonary vein. Resting A-VO$_2$ difference is 4–5 mL/100 mL blood, and during training it may increase to 16–18 mL/100 mL. There is a difference between trained and untrained athletes as well. Upon endurance training, oxygen content in venous blood is lower, indicating that tissues uptake oxygen more efficiently. The underlying cause behind the latter is the increased vascularization following endurance training. Saturation of hemoglobin, which is an important factor determining A-VO$_2$ difference, does not change during training; however, it may decrease at high altitude because of the lower atmospheric pressure. In some cases, hematocrit level may increase due to mobilization of RBC from bone marrow and spleen to improve oxygen carrier capacity.

5.2.3 Peripheral Blood Supply and Vascularization as One of the Main Factors Determining VO₂max

Following oxygen uptake in the lungs, oxygen needs to reach the tissues in muscles during physical training. The more dense the capillary network in the muscle is, the more efficient the oxygen consumption is. As an analogy from everyday life, the more trucks deliver stocks to a store and can park close to the entrance, the faster the stocks may be placed on the shelves in the store. Types of training and environment factors such as high altitude alter and increase vascularization significantly. In most cells and so in the muscles there are proteins that able to sense oxygen content, such as hypoxia inducible factor-1 (HIF-1α). During physical exercise, oxygen demand is higher in the muscle compared to the supply, resulting in hypoxia, which is sensed by HIF-1α. This induces vascular endothelial growth factor (VEGF), leading to new vessel formation. Other targets of HIF-1α are genes that help the adaptation to lower oxygen content; these genes encode proteins that are involved in anaerobic metabolic processes. An undeveloped capillary network is a limiting factor in oxygen and nutrient supply; abundant vascularization improves this. Vascularization can be promoted by well-designed training (Fig. 5.7).

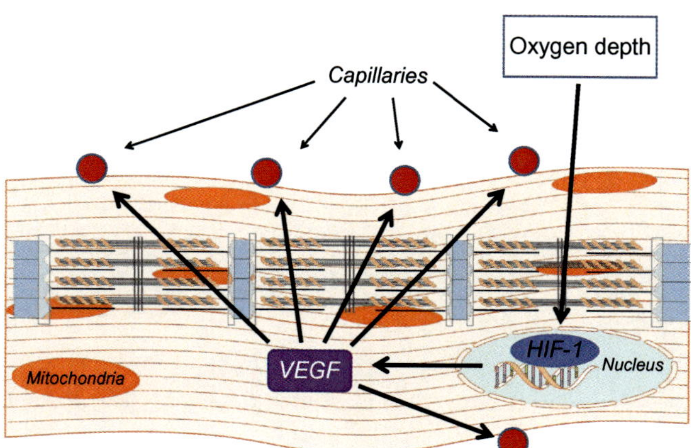

Fig. 5.7 Hypoxia and vascularization. Hypoxia or physical training activates hypoxia inducible factor-1 (HIF-1α) resulting in abundant vascularization.

5.3 MITOCHONDRIAL CONTENT AND ENZYMATIC ACTIVITY AS ONE OF THE MAIN FACTORS DETERMINING VO$_2$MAX

Untrained athletes show improved performance following frequent endurance training, which is a result of adaptive responses in the muscle, and not mainly the increase of VO$_2$max (Holloszy and Coyle, 1984). Skeletal muscle and mitochondrial functions are associated closely to endurance performance, and some authors emphasize the primacy of the muscle (Timmons, 2011). Indeed, endurance is determined by many factors, and one of VO$_2$max's limiting factors is skeletal muscle. Mitochondrial density in the muscle and mitochondrial enzyme activity and quantity can be increased by physical training. Mitochondria are 0.5–1.0 μm cellular organelles; in the muscle they are subsarcolemmal or intermyofibrillar depending on their localization. Subsarcolemmal mitochondria are involved mainly in ATP production necessary for Na-K pump function, while ATP produced by intermyofibrillar mitochondria is consumed mainly by contraction. It has been known since the 1960s that the number of mitochondria can be increased by physical training. This is one of the most important adaptations to endurance training. ATP is produced by mitochondria, thus the more mitochondria a cell contains, the more efficiently the work may be done. Slow-twitch muscle fibers are red in color due to the iron content of myoglobin and mitochondria. Excellent endurance requires a significant amount of slow-twitch muscle fibers.

What are the mechanisms that result in an increased number of mitochondria? One factor is increased ATP demand during training. ATP metabolism results in AMP production, which activates adenosine monophosphate protein kinase (AMPK), which in turn stimulates mitochondrial biogenesis, division, and fusions. AMPK plays a role in anaerobic/aerobic transition (Barnes et al., 2005), from glycolysis to oxidative metabolic processes, and it also means that primarily aerobic and mixed (anaerobic/aerobic) trainings activate mitochondrial biogenesis via AMPK. However, experimental data show that 20–30 s. maximal intensity training results in increased number of mitochondria (Gibala et al., 2009), which is not surprising since maximal intensity training induces hypoxia, thus an ADP+ADP=ATP+AMP process frequently takes place. Thus, AMP stimulates AMPK and mitochondrial biogenesis in order to increase ATP supply during physical training.

During contraction, Ca^{2+} efflux from the sarcoplasmic reticulum also stimulates mitochondrial biogenesis. The most important stimulator of mitochondrial biogenesis is PGC-1 alpha transcriptional coactivator, which can be induced more efficiently by high intensity training (Tadaishi et al., 2011).

In response to endurance training, the activity of several oxidative metabolic enzymes may double such as succinate dehydrogenase, NADH-dehydrogenase, NADH-, cytochrome c-reductase, and cytochrome c-oxidase (Holloszy and Coyle, 1984). The activity and quantity of additional enzymes involved in fatty acid transport, beta-oxidation, and citrate cycle, are also increased following physical training. Citrate synthase's activity is a good marker of skeletal muscle's adaptation since it reacts dynamically to training and longer cessation from training (detraining). Detraining due to injuries influences physical performance. It has been demonstrated that following 15 weeks of training, increased enzymatic activity returned to base level after 4–5 weeks of detraining (Henriksson and Reitman, 1976). Many years of endurance training (6–20 years) followed by 12 weeks of detraining decreased enzymatic activity significantly; however, it remained 40% above the control value (Chi et al., 1983). The activity of these aerobic-specific enzymes decreases in fast-twitch muscle fibers to a lower extent compared to slow–twitch muscle fibers. This may indicate that fast-twitch fibers shifted to slow–twitch fiber type. This emphasizes the importance of well-designed training, since optimal training intensity influences performance in particular sport events. For instance, football players would only receive low intensity training; their fast-twitch muscle fibers regardless of their dominance would not be as efficient as they are needed for sprinting and jumps.

The adaptation or genetic characteristics of the abovementioned factors are all determinants of VO_2max, which is one of the most important markers of endurance. However, performance in an endurance sport is determined by other factors in addition to VO_2max. A certain VO_2max determines whether an individual can be successful in certain sports; however, this is only a prerequisite and not a warranty for excellent performance. The characteristics of skeletal muscle are the additional factor to VO_2max in outstanding performance.

5.4 EFFECT OF METABOLIC PROCESSES ON ENDURANCE

Endurance training induces adaptation in metabolic processes in addition to the cardiovascular system and skeletal muscle. ATP demand during training, which is intensity-dependent, determines the type of metabolic process, the percentile of energy sources needed for ATP production such as creatine-phosphate, adenylate kinase reaction, anaerobic and aerobic glycolysis, and fat oxidation. ATP hydrolysis is intensity-dependent, while synthesis depends on the number of mitochondria and enzymes mentioned above.

Cellular ATP, ADP, AMP, Pi, and the adenylate cyclase side product, ammonia, influence metabolic processes and are responsible for the lower lactic acid production in trained athletes compared to untrained ones. Two ADP make up one ATP, and the rest of AMP is metabolized, resulting in ammonia generation. Accumulated ammonia adversely alters physiological functions. Pi is the product of ATP metabolism. These side products exert their influences through cellular signal pathways, and play a role in the development of fatigue.

In response to endurance training, the activity and the amount of LDH subunits increase, the latter catalyzing lactic acid conversion to pyruvate. Metabolic adaptation to endurance training manifests in decreases carbohydrate consumption and elevated fat metabolism. Free fatty acid concentration in blood is higher during training in trained athletes. Fat provides more energy than carbohydrates; however, its metabolism demands more oxygen, thus it can be used efficiently only during low or moderate intensity exercise. Trained individuals with higher VO_2max are able to use fat efficiently as an energy source at higher intervals of intensity; however, high intensity training recruits carbohydrates for ATP production. At moderate intensity, trained athletes are able to mobilize and metabolize fat more quickly than untrained ones. From a metabolic point of view, the highest peak of intensity is where energy production still occurs through aerobic pathways. This is a good indicator of endurance.

Energy sources are abundant such as fat, carbohydrate, and proteins. Even alcohol is a quickly employable energy source, and although it cannot be accumulated, it is used up before other energy sources. Although it may be an attractive energy source, in high concentration it is toxic. Lactic acid can also be used as an energy source (Fig. 5.8).

Energy consumption during resting is made up of two-thirds fat breakdown and one-third carbohydrate breakdown. An hour-long aerobic training session consumes energy from fat 50%–60%, carbohydrate 40%–50%, and protein 1%. Aerobic training lasting 4–5 h increases protein consumption up to 5% of total energy sources, and fat breakdown can be 80%–90% of total energy sources, while carbohydrate is about 10% (Hood and Terjung, 1990; Jeukendrup, 2002, 2003). Trained individuals cover energy 50%–50% from fat and carbohydrate during 20–30 min. Training, and untrained ones need to train for longer to achieve this rate. Exercise intensity in endurance is defined as a percentage of VO_2max. During VO_2max 20% intensity, 60% of total energy derives from fat, whereas it decreases to 40% when VO_2max is 50%. Considering

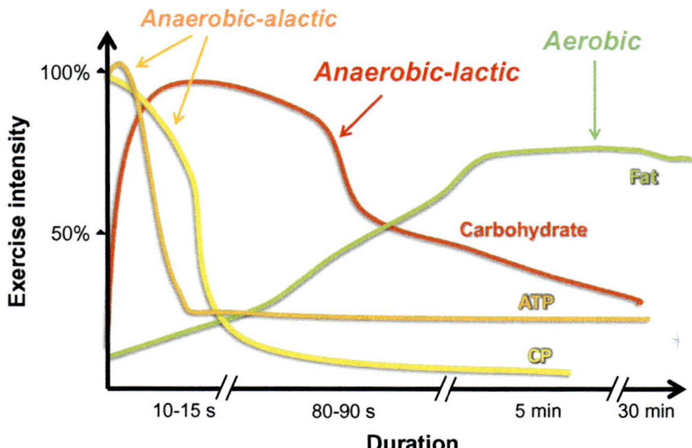

Fig. 5.8 Energy sources and energy production. Different energy sources used at different training intensities. ATP is the main source of energy at a cellular level, and it needs to be reproduced. The type of energy source (CP, carbohydrate, fat) and its form of breakdown (aerobic or anaerobic) are intensity-dependent.

that the higher intensity training's total energy consumption is higher by approximately 2.5 times, its total fat consumption is 33% higher than low intensity training.

As intensity of training increases and duration decreases, carbohydrate will be the main energy source, and lactic acid production increases (Gollnick et al., 1973). Resting and aerobic training also produce lactic acid in the muscle and the brain since it has a function during glycolysis (as a proton acceptor) (Pilarski et al., 2009); however, it is used to describe anaerobic metabolism, and it has a role in fatigue (Brooks, 1986). Resting lactic acid concentration in the blood is 1–2 mmol/L, which may increase up to 20–22 mmol/L in professional athletes during maximal loading. Elevated lactic acid concentration inhibits fat metabolism (Turcotte, 2000). During high intensity training (VO$_2$max over 100%), anaerobic processes dominate in ATP synthesis. ATP is produced from carbohydrates in the anaerobic lactic process, while in anaerobic alactic processes, ATP is produced from creatine phosphate (CP) and adenylate cyclase reaction (ADP+ADP=ATP+AMP). CP consumption is higher in high intensity training compared to low intensity training, even if blood pH is similar at the end of training (Forbes et al., 2009). CP recovery is quick and needs only minutes, whereas lactic acid elimination needs more time (about 30 min), and regeneration of glycogen stores need even more time (a couple of days) (Table 3.2).

5.5 LACTATE THRESHOLD

The purpose of endurance training is the maximization of both intensity and length of aerobic energy production and minimization of anaerobic energy generation. The physiological aim is efficiency; one single molecule of glucose provides 18 times more ATP through aerobic metabolism than anaerobic breakdown, and also lactic acid accumulation and its adverse effects can be avoided (effects of lactic acid are discussed in detail in the "R" Extra section of this chapter). Lactic acid is a well-known marker of anaerobic metabolism, which can be measured from one drop of blood taken from a fingerprint or earlobe. Lactic acid is also produced during aerobic metabolism, for instance, during sleeping; however, during loading its concentration increases exponentially in the blood (Fig. 5.9). The lactic acid elevation curve shifts to the right in trained athletes.

Lactic acid concentration during exercise is determined by several factors such as saturation of glycogen stores in the muscle (less glycogen results in less lactic acid; however, it is not adaptation and this low energy state results in moderate performance) (Yoshida, 1984), muscle fibers composition, vascularization, and the activity of oxidative enzymes. In the late

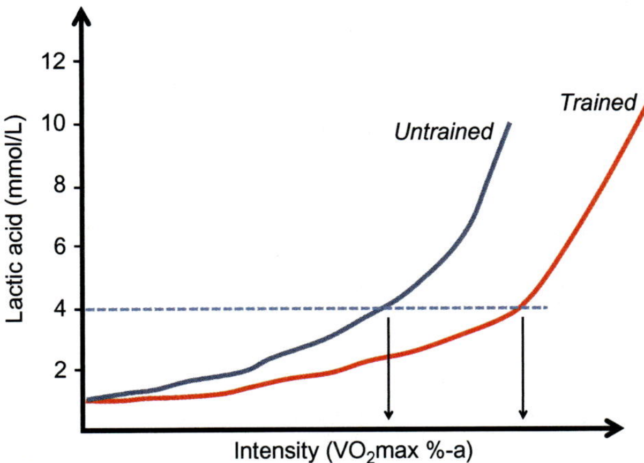

Fig. 5.9 Lactate threshold. Lactic acid production increases in a nonlinear manner during increasing training intensity; at a certain point it becomes a steeper curve. Lactic acid production and elimination are regulated during aerobic metabolism, not exceeding 4 mmol/L. Under anaerobic, hypoxic conditions, lactic acid production outgrows elimination capacity, and increases exponentially in muscle and blood. Higher exercise intensity can be seen in trained athletes when they reach the lactic acid threshold, the point from where lactate production elevates steeply. The intensity where athletes reach the anaerobic threshold is a good indicator of aerobic fitness, and this can be improved by training.

1970s, Kindermann et al. (1979) introduced the anaerobic threshold to measure aerobic–anaerobic transition. In this model, there are two values; one is aerobic threshold. Lactic acid level is above resting lactic acid concentration, and this level allows long-lasting steady-state training, where energy production is fulfilled by aerobic ATP synthesis. This training improves cardiovascular conditions, which will be further discussed in the section on endurance training methods. Higher intensity recruits anaerobic metabolic processes accompanied by elevated lactic acid production; however, this lactic acid can be eliminated efficiently in the muscle and liver. Further increase of training intensity results in an aerobic–anaerobic transition, where lactic acid production and elimination are in equilibrium; this is called the anaerobic threshold. At this point, blood lactate concentration is about 4 mmol/L; however, individual differences show 2–10 mmol/L concentration (Noordhof et al., 2010) since it is determined by fitness, and also by genetic factors. Above the anaerobic threshold, lactic acid production increases at a higher rate than it could be eliminated, thus it shows an exponential increase in the blood (Fig. 5.10).

Increasing intensity recruits anaerobic processes gradually; however, lactic acid production is nonlinear, and from a certain point it suddenly increases at a high rate. Practically, setting the optimal training intensity is

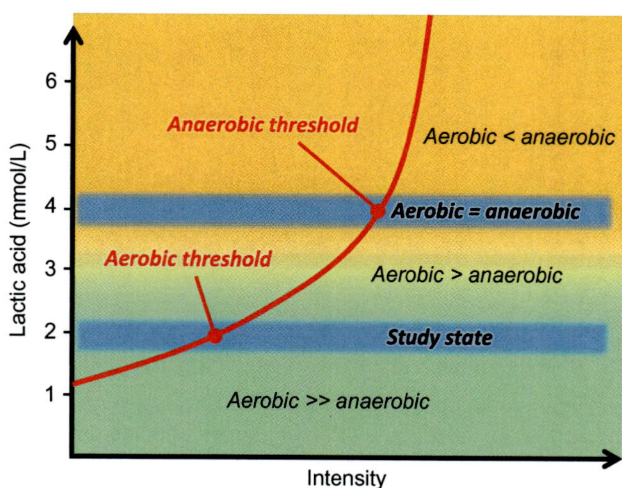

Fig. 5.10 Training intensity determining metabolic processes. Energy demand is fulfilled from many sources. Low intensity training induces aerobic metabolic processes, where energy production is fulfilled by aerobic ATP synthesis. Higher intensity results in increased rate of anaerobic energy production, and above the anaerobic threshold, it is the only way to fulfill energy demand.

based on the anaerobic threshold, and with improved fitness, training intensity can be elevated. Above a certain VO_2max (about 55–60 mL/kg/min in females, and 65–70 mL/kg/min in males), the velocity of reaching the anaerobic threshold is a better marker of endurance (Coyle et al., 1991). There is a close association between performance around the anaerobic threshold and the achievement on 5–42 km (Jones and Doust, 1998). Trained athletes are able to supply energy demand via aerobic metabolism at higher heart rates, and the heart rate at the anaerobic threshold can be 80%–90% of the maximal value, whereas that value is 65%–75% in untrained individuals (Bosquet et al., 2002). Thus, the goal of endurance training is to enhance the aerobic zone in higher training intensity.

The lactate threshold is also used in the literature in addition to the anaerobic threshold, and it indicates a lactate concentration of 4 mmol/L in the blood. Another term, OBLA (onset of blood lactate accumulation), also appears. These terms all refer to the aerobic–anaerobic transition point (Noordhof et al., 2010). Determining training intensity at the anaerobic threshold is important, as optimal loading can be adjusted by keeping the heart rate around this threshold. Measuring heart rate during training provides information to ensure optimal intensity.

5.6 ENDURANCE TRAINING METHODS

Endurance training focuses on the advancement of the limiting factors to improve physical performance discussed above. Specific adaptations to different types of training need careful design of exercises. For instance, different training is needed to improve the stroke volume of the heart, and to increase lactic acid tolerance. Endurance can be developed easily; however, genetic factors also play a role in the process. Twin studies showed VO_2max higher similarities in identical than fraternal twins (Bouchard et al., 1986). Adaptation following 15 weeks of endurance training showed similarities in identical twins compared to fraternal ones. And analysis of individuals with different genetics showed different adaptation to 15 weeks of training; thus genetic factors influence adaptation responses to physical training (Hamel et al., 1986).

In terms of endurance, VO_2max influences quality of life, and low VO_2max limits physical performance. Endurance is a complex physiological ability, which is separated into categories such as aerobic, anaerobic lactic, and anaerobic alactic based on metabolic properties. The recommended diet for endurance training will be discussed in Chapter 8, and genetic factors will be described in Chapter 11.

5.6.1 Aerobic Endurance Training Methods

Aerobic endurance training is sometimes interchanged with aerobic capacity training. It is also an important part in endurance training; however, training intensity influences the heart and the skeletal muscle differently, the two main determining factors in endurance. We address adaptation following acute (following a single training) and chronic (frequent trainings) loadings.

Continuous, Long-Lasting Endurance Training

This is the most popular endurance training for both beginners and professional athletes. A training at VO_2max 50%–70% longer than 30 min. Causes moderate stress to the body and has beneficial effects on body weight, cardiovascular prevention, and improvement of maximal aerobic performance.

An acute single workout for a couple of hours may result in muscle and joint pain regardless of moderate intensity. Long-lasting loading causes dehydration due to sweating, thus liquid supplementation is important during training. During this steady-state loading, fatty acid mobilization can be significant. This type of training is sometimes called a fat burning workout. This training also mobilizes glycogen stores of the muscle.

This training improves VO_2max due to its cardiovascular effects since it increases the size of the left ventricle and also stroke volume. In response to long-lasting low intensity workouts, the muscle stores more glycogen and the activity of enzymes involved in fat metabolism are also elevated. This training also improves tolerance to monotony. This method is recommended at every age, for young and old, for beginners and professional athletes.

Fartlek Speed Play Training

This is also a continuous training blending with interval training with different intensity instructed by the trainer or by the athlete themselves. This method primarily improves aerobic endurance; the average intensity of the training is about VO_2max 60%–80% and lasts between 30 min and 1.5 h. Acute workout cause increased lactic acid production; however, it is around the anaerobic threshold. Depending on the duration, joint pain may occur, and glycogen stores reduce. Adaptation to lactic acid is induced by frequent trainings.

Speed play training improves cardiovascular status and VO_2max efficiently, and increases glycogen storing capacity. Varying intensity improves pacing skills, which are important when selecting the optimal speed and pace. The activity of enzymes involved in carbohydrate metabolism is also elevated.

This method is recommended for the young; it is safe only for well-trained elderly. This training with its varying intensity mimics the pacing of ball games and martial arts, thus athletes of these sports can also benefit.

Aerobic Interval Training

This method consists of short, intense efforts followed by active or passive recovery. Intensity still falls in the aerobic zone, VO_2max 70%–80%, duration of intervals are usually over 5 min. During the recovery period, heart rate should not fall below 100–120 beat/min in adult athletes. This intensity induces also anaerobic metabolism; however, lactic acid can be eliminated efficiently.

During an acute workout, the rate of fat mobilization decreases, and primarily carbohydrates provide the energy for ATP generation. The high heart rate cannot allow the left ventricle to fill up, thus unlike during long-lasting low intensity trainings, stroke volume does not increase. Adaptation responses include increased number of mitochondria and enzyme activity involved in oxidative metabolism; vascularization of the musculature is also significant.

This method is recommended for all ages, especially for 12- to 16-year-olds to establish an excellent musculature and cardiovascular fitness necessary for outstanding endurance later.

5.6.2 Anaerobic Endurance Training Methods

Anaerobic endurance training is beneficial for athletes whose sports set a high intensity and whose bodies need to utilize anaerobic metabolic processes. High intensity interval training improves both aerobic and anaerobic capacity, e.g., mitochondrial content of the muscle (Weston et al., 2014). Anaerobic training usually involves short, intense efforts. It depletes only CP, thus regeneration is quick following exercises, and this makes it very popular among professional athletes.

Anaerobic Threshold Interval Training

Training intensity is around the anaerobic threshold, around VO_2max 70%–90%, duration 3–8 min, and 5–15 serials. Fatty acid breakdown is not significant in this intensity zone, and not more than 70% of total energy consumption rises from aerobic carbohydrate metabolism. The main goal of this training is to increase the anaerobic threshold, thus both aerobic and anaerobic ATP production are important. During recovery, heart rate decreases to 120 beat/min. Lactic acid concentration increases in the muscle and the blood, which cannot be eliminated as efficiently, therefore its concentration is 3–6 mmol/L in the blood. It induces fatigue and pain, which is caused by lactic acid.

Adaptation occurs mainly in skeletal muscle in this intensity zone, and it also improves arterio-venous oxygen difference. The elevated AMP results in mitochondrial biogenesis, and activity of both aerobic and anaerobic enzymes increases. Intensity at anaerobic threshold can be improved by this method. This method is recommended following puberty.

Anaerobic Interval Training

Training intensity is at VO_2max 90%–120%, duration 20 s to 3 min. ATP production rises mainly from anaerobic metabolism, and lactic acid concentration is elevated; it may go beyond 20 mmol/L in well-trained athletes. Recovery lasts until heart rate reaches 120 beat/min. The heart rate is close to maximal during exercising, thus the left ventricle does not have enough time to fill up, so this method does not influence stroke volume. High lactic acid level induces fatigue and pain.

Although mitochondrial biogenesis occurs due to high AMP level, increased activity of enzymes involved in anaerobic metabolism is more significant. Fast-twitch muscle fibers are recruited due to the high intensity, and hypoxia activates HIF-1α, which induces vascularization in the muscle. This method results in increased resistance to lactic acid, and A-VO_2 difference improves. This method is recommended for junior and adult athletes, and it can also be included in training plans of younger athletes to some extent.

Mini Interval Training

During maximal intensity at VO_2max 120%–150%, anaerobic alactic and anaerobic lactic processes provide energy. Duration of exercise is 4–20 s. Lactic acid production is significant. During recovery, the individual should wait until his or her heart rate reaches 120 beat/min. This type of training induces significant fatigue and pain.

This training depletes CP significantly; however, it increases resistance to lactic acid, and improves enzyme activity involved in lactic acid elimination and anaerobic metabolism. CP recovery improves and CP level will be higher in the muscles (Larsen et al., 2013). High intensity induces cross-sectional area growth due to anabolic hormone secretion and fast-twitch muscle recruitment. Mini interval training is beneficial for ball game sports, and for professional athletes. However, it can also be included in training plans of younger athletes.

5.6.3 High-Altitude Training

High-altitude training at 2000–2500 m is the most efficient for training, whereas higher altitudes (e.g., 3500–4000 m) have adverse health effects, and lead to high-altitude sickness. The main factor at high altitudes is the lower atmospheric pressure, which results in impaired gas exchange and impaired hemoglobin saturation. Staying at high altitudes has both beneficial and adverse effects; above 3500 m it has more adverse effects.

It must be noted that native people living at high altitudes are genetically adapted to such environment. Acute effects of high-altitude life include increased respiratory rate, increased resting heart rate, decreased maximal heart rate, decreased VO_2max, increased resting lactic acid level, and impaired endurance. High-altitude aerobic type competitions result in weaker performance compared to those at sea-level. Lower resistance due to thinner air may lead to better performance in sport categories where speed is in the focus. Preparation for competitions at high altitudes needs high-altitude training, whereas beneficial effects of high-altitude training on performance at sea-level is less evident. Effects of high-altitude training on speed performance and regeneration were examined by eight cyclists at 585 and 2100 m. Workouts lasted 3x6x15 sec at maximal intensity, and rates of loading and resting periods were 1:3 (15 s:45 s), 1:2 (15 s:30 s), and 1:1 (15 s:15 s). Performance at a high altitude decreased by 5%, when the loading-resting rate was 1:3, and decreased by 10% when the loading-resting rate was 1:1 (Brosnan et al., 2000). Thus, speed performance is adversely affected by high-altitude and increased the necessary time for regeneration. Six weeks of aerobic type training at high altitudes does not show differences compared to training at sea-level (Robach et al., 2014).

High-altitude training results in a higher count of RBC, increased A/VO_2 difference, effective oxygen utilization, denser vascularization, and an increased amount of mitochondria. Thus, high-altitude training may be beneficial in aerobic endurance performance; however, it does not always lead to better performance at sea-level. Training intensity is lower at high altitudes due to impaired gas exchange and VO_2max (Fig. 5.11).

At an altitude of 6000 m, VO_2max decreases to 50% of sea-level value, thus fewer fast-twitch muscle fibers are recruited at submaximal or maximal efforts. To solve this issue, the "sleep high train low" method has been developed. Athletes sleep in altitude tents resulting in adaptation of RBC, A/VO_2 difference, while training at sea-level allows high intensity exercise. This method has been widely investigated, and shows better efficiency compared to high-altitude training (Millet et al., 2010). It also has a further developed

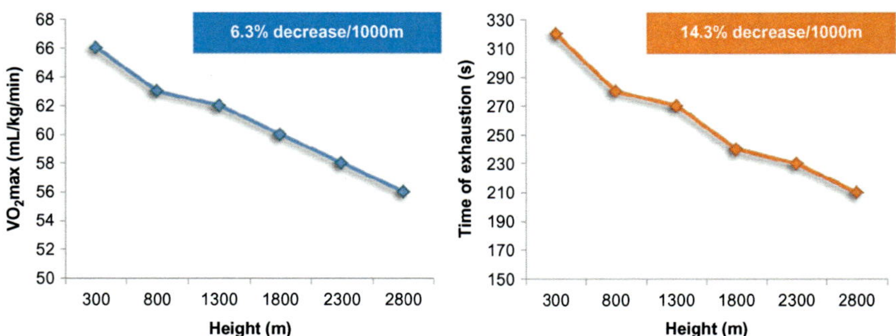

Fig. 5.11 High-altitude training. A high altitude results in decreased VO$_2$max, and time period until exhaustion. Thus, a high altitude leads to impaired aerobic performance. Regardless, high-altitude training with a mask or altitude tent is popular to improve endurance.

variation, which comprises of high intensity interval training at sea-level (Czuba et al., 2014). This method seems to be effective in improving both aerobic and anaerobic endurance.

A high altitude also influences metabolic processes due to low oxygen concentration, fat oxidation decreases, and carbohydrate metabolism is more important at high altitude than fat metabolism. Thus, a diet high in carbohydrate content is more beneficial at high altitudes. Iron as a component of hemoglobin also needs to be available in sufficient amounts.

5.7 "R" EXTRA

Aerobic endurance, sometimes just called endurance, is necessary in every sport. Implementing endurance training is most challenging in ball sports since aerobic, anaerobic lactic, and alactic processes present in combination, and there are significant differences in players' positions. For instance, the average loading intensity in UK premier league football is about VO$_2$max 70%; however, the best athletes complete 150–250 short, high intensity movements. This results in phosphate depletion and increased lactic acid level due to anaerobic lactic processes (Bangsbo et al., 2007). Glycogen content of the muscle may decrease by 40%–90%, which emphasizes the beneficial effects of carbohydrate loading.

VO$_2$max of male champions in aerobic endurance sports is about 70–85 mL/kg/min (Costill et al., 1973), while this value is 10% less in females due to smaller mass muscle, higher fat content, and fewer RBC. Training intensity is 100% VO$_2$max in 5 km track and field running, 90%–100% in 10 km running, and 75%–80% in marathon running (Bassett and Howley, 2000).

ATP production in 5 km running rises from anaerobic processes at a rate of 10%–20%, while in marathon running this rate is 1%–5% (Bangsbo et al., 1993). Intensity value associated with the anaerobic threshold shows significant individual differences even among individuals with similar VO_2max, which is a result of differences in muscle vascularization. Denser vascularization results in higher velocity value at the anaerobic threshold (Coyle, 1995). Efficiency is very important in endurance performance, which is determined by slow-twitch muscle fibers (Coyle, 1995), since these fibers works more efficiently than fast-twitch muscle fibers.

VO_2max reaches its peak at the age of 35; after this age it declines regardless of frequent training, and decreases significantly over 60. This applies to senior master athletes as well. The underlying cause behind decreasing VO_2max is the heart, since maximal heart rate, stroke volume, and A/VO_2 difference declines with age. Efficiency of movements does not change; however, the intensity level associated with anaerobic threshold decreases (Tanaka and Seals, 2008). Regardless, endurance improves well in response to training also in the elderly. Comparing first-place scores of the first modern Olympics (1896) in 1500 m running (4:33.2) and marathon running (2:58:50) to scores achieved by the elderly in these days: a 60-year-old athlete runs 1500 m in 4:27.7, while a 73-year-old athlete runs the marathon in 2:54:05. These data show nicely that endurance improves well in response to training.

Low intensity long-lasting training or aerobic interval training were thought to improve VO_2max the most efficiently; however, data show that 6- to 9-week interval training with 90%–120% intensity also increases VO_2max and mitochondrial biogenesis (Sloth et al., 2013). This method shortens the length of a training session. The main difference between aerobic long-lasting training and the high intensity training is that carbohydrate breakdown dominates over fat oxidation in short high intensity trainings. Both training methods improve VO_2max efficiently; however, cardiovascular and metabolic adaptation are different. The continuous low intensity aerobic method is recommended for everybody, while high intensity training is recommended mainly for young and professional athletes.

High-altitude hiking and climbing is very popular; however, it has some risks. This type of training is not recommended above 3500 m (11,000 ft), and extreme high risks are present above 8000 m (26,000 ft). Altitude sickness is a complex phenomenon, and sensitive individuals may develop symptoms at 4000 m (13,000 ft), which are not definitely correlated with fitness. High-altitude sickness symptoms include headache due to altered blood-brain barrier permeability and cerebral edema. Pulmonary edema is also a potentially lethal condition. Climbing above 4000 m requires a couple of days' acclimatization first at lower altitudes, around 2000–3000 m. To reach 6000 or 8000 m, one needs longer adaptation at different altitudes. The main sensor of low oxygen in the body is HIF-1α, which stimulates EPO production in the kidneys and consequential increase in RBC count, and vascularization via VEGF. High altitudes stimulate free radical production and oxidative stress (Dosek et al., 2007).

5.8 SUMMARY

Endurance improves well in response to training; however, genetic factors also determine success. Endurance can be divided into categories such as aerobic, anaerobic lactic, and anaerobic alactic due to differences in metabolic processes. The main energy sources for ATP synthesis are fat and carbohydrates in aerobic, carbohydrates in anaerobic lactic, and CP in anaerobic alactic processes. Aerobic metabolism provides more energy, thus the longer that aerobic metabolism can provide energy and burn fat, the more successful an endurance athlete can be. Therefore, VO_2max mainly determines endurance. At a young age males have an average VO_2max 50 mL/kg/min, professional athletes have 70–85 mL/kg/min, and females have values 10% less. VO_2max is determined by cardiac output, RBC count, A/VO_2 difference, vascularization of skeletal muscle, muscle fibers type, and the amount of mitochondria and the activity of oxidative enzymes.

Intensity associated with anaerobic threshold is another important factor that determines endurance. It marks an intensity level where aerobic and anaerobic processes are in equilibrium; over this point, anaerobic processes dominate. This point is at an intensity of 20%–30% higher (VO_2max) in professional athletes compared to less trained individuals. The main goal of endurance training is to increase the anaerobic threshold, because this improves training efficiency. Determining factors of VO_2max react differently to different training intensity and duration, and genetic factors also determine the VO_2max of an individual.

TEST QUESTIONS

1. What factors determine VO_2max?
2. What training methods influence these factors?
3. Draw a diagram of energy sources and energy production in relation to training intensity.

4. What is the anaerobic threshold, and what is its importance in endurance training?
5. What are the characteristics of adaptation to endurance training?
6. What training methods are efficient to improve heart functions and how do they work?
7. What training methods would you recommend to improve oxidative functions of mitochondria in skeletal muscles?
8. How would you improve resistance to lactic acid and how does it work?
9. What type of endurance training would you recommend a football player and a 1 min. Swimmer?
10. Why is high-altitude training not always useful to improve endurance?
11. What type of training would you recommend to improve health and how does it work?

BIBLIOGRAPHY

Baggish, A.L., Wang, F., Weiner, R.B., Elinoff, J.M., Tournoux, F., Boland, A., Picard, M.H., Hutter Jr., A.M., Wood, M.J., 2008. Training-specific changes in cardiac structure and function: a prospective and longitudinal assessment of competitive athletes. J. Appl. Physiol. 104, 1121–1128.

Bangsbo, J., Michalsik, L., Petersen, A., 1993. Accumulated O_2 deficit during intense exercise and muscle characteristics of elite athletes. Int. J. Sports Med. 14, 207–213.

Bangsbo, J., Iaia, F.M., Krustrup, P., 2007. Metabolic response and fatigue in soccer. Int. J. Sports Physiol. Perform. 2, 111–127.

Barnes, B.R., Glund, S., Long, Y.C., Hjalm, G., Andersson, L., Zierath, J.R., 2005. 5′-AMP-activated protein kinase regulates skeletal muscle glycogen content and ergogenics. FASEB J. 19, 773–779.

Bassett Jr., D.R., Howley, E.T., 2000. Limiting factors for maximum oxygen uptake and determinants of endurance performance. Med. Sci. Sports Exerc. 32, 70–84.

Bosquet, L., Leger, L., Legros, P., 2002. Methods to determine aerobic endurance. Sports Med. 32, 675–700.

Bouchard, C., Lesage, R., Lortie, G., Simoneau, J.A., Hamel, P., Boulay, M.R., Perusse, L., Theriault, G., Leblanc, C., 1986. Aerobic performance in brothers, dizygotic and monozygotic twins. Med. Sci. Sports Exerc. 18, 639–646.

Brooks, G.A., 1986. Lactate production under fully aerobic conditions: the lactate shuttle during rest and exercise. Fed. Proc. 45, 2924–2929.

Brosnan, M.J., Martin, D.T., Hahn, A.G., Gore, C.J., Hawley, J.A., 2000. Impaired interval exercise responses in elite female cyclists at moderate simulated altitude. J. Appl. Physiol. 89, 1819–1824.

Chi, M.M., Hintz, C.S., Coyle, E.F., Martin 3rd, W.H., Ivy, J.L., Nemeth, P.M., Holloszy, J.O., Lowry, O.H., 1983. Effects of detraining on enzymes of energy metabolism in individual human muscle fibers. Am. J. Phys. 244, C276–C287.

Costill, D.L., Thomason, H., Roberts, E., 1973. Fractional utilization of the aerobic capacity during distance running. Med. Sci. Sports 5, 248–252.

Coyle, E.F., Feltner, M.E., Kautz, S.A., Hamilton, M.T., Montain, S.J., Baylor, A.M., Abraham, L.D., Petrek, G.W., 1991. Physiological and biomechanical factors associated with elite endurance cycling performance. Med. Sci. Sports Exerc. 23, 93–107.

Coyle, E.F., 1995. Integration of the physiological factors determining endurance performance ability. Exerc. Sport Sci. Rev. 23, 25–63.

Czuba, M., Maszczyk, A., Gerasimuk, D., Roczniok, R., Fidos-Czuba, O., Zajac, A., Golas, A., Mostowik, A., Langfort, J., 2014. The effects of hypobaric hypoxia on erythropoiesis, maximal oxygen uptake and energy cost of exercise under normoxia in elite biathletes. J. Sports Sci. Med. 13, 912–920.

Dosek, A., Ohno, H., Acs, Z., Taylor, A.W., Radak, Z., 2007. High altitude and oxidative stress. Respir. Physiol. Neurobiol. 158, 128–131.

Duncker, D.J., Bache, R.J., 2008. Regulation of coronary blood flow during exercise. Physiol. Rev. 88, 1009–1086.

Durakovic, Z., Misigoj-Durakovic, M., Medved, R., Skavic, J., Torovic, N., 2002. Sudden death due to physical exercise in the elderly. Coll. Antropol. 26, 239–243.

Forbes, S.C., Paganini, A.T., Slade, J.M., Towse, T.F., Meyer, R.A., 2009. Phosphocreatine recovery kinetics following low- and high-intensity exercise in human triceps surae and rat posterior hindlimb muscles. Am. J. Physiol. Regul. Integr. Comp. Physiol. 296, R161–170.

Gibala, M.J., McGee, S.L., Garnham, A.P., Howlett, K.F., Snow, R.J., Hargreaves, M., 2009. Brief intense interval exercise activates AMPK and p38 MAPK signaling and increases the expression of PGC-1alpha in human skeletal muscle. J. Appl. Physiol. 106, 929–934.

Gollnick, P.D., Armstrong, R.B., Saubert, C.W., Sembrowich, W.L., Shepherd, R.E., Saltin, B., 1973. Glycogen depletion patterns in human skeletal muscle fibers during prolonged work. Pflugers Arch. 344, 1–12.

Hamel, P., Simoneau, J.A., Lortie, G., Boulay, M.R., Bouchard, C., 1986. Heredity and muscle adaptation to endurance training. Med. Sci. Sports Exerc. 18, 690–696.

Henriksson, J., Reitman, J.S., 1976. Quantitative measures of enzyme activities in type I and type II muscle fibres of man after training. Acta Physiol. Scand. 97, 392–397.

Hill, A.V., Lupton, H., 1923. Muscular exercise, lactic acid, and the supply and utilization of oxygen. QJM 16, 135–171.

Holloszy, J.O., Coyle, E.F., 1984. Adaptations of skeletal muscle to endurance exercise and their metabolic consequences. J. Appl. Physiol. Respir. Environ. Exerc. Physiol. 56, 831–838.

Hood, D.A., Terjung, R.L., 1990. Amino acid metabolism during exercise and following endurance training. Sports Med. 9, 23–35.

Jeukendrup, A.E., 2002. Regulation of fat metabolism in skeletal muscle. Ann. N.Y. Acad. Sci. 967, 217–235.

Jeukendrup, A.E., 2003. Modulation of carbohydrate and fat utilization by diet, exercise and environment. Biochem. Soc. Trans. 31, 1270–1273.

Jones, A.M., Doust, J.H., 1998. The validity of the lactate minimum test for determination of the maximal lactate steady state. Med. Sci. Sports Exerc. 30, 1304–1313.

Kindermann, W., Simon, G., Keul, J., 1979. The significance of the aerobic-anaerobic transition for the determination of work load intensities during endurance training. Eur. J. Appl. Physiol. Occup. Physiol. 42, 25–34.

Larsen, R.G., Befroy, D.E., Kent-Braun, J.A., 2013. High-intensity interval training increases in vivo oxidative capacity with no effect on P(i)→ATP rate in resting human muscle. Am. J. Physiol. Regul. Integr. Comp. Physiol. 304, R333–342.

Lauschke, J., Maisch, B., 2009. Athlete's heart or hypertrophic cardiomyopathy? Clin. Res. Cardiol. 98, 80–88.

Marliss, E.B., Vranic, M., 2002. Intense exercise has unique effects on both insulin release and its roles in glucoregulation: implications for diabetes. Diabetes 51 (Suppl 1), S271–283.

Millet, G.P., Roels, B., Schmitt, L., Woorons, X., Richalet, J.P., 2010. Combining hypoxic methods for peak performance. Sports Med. 40, 1–25.

Noordhof, D.A., de Koning, J.J., Foster, C., 2010. The maximal accumulated oxygen deficit method: a valid and reliable measure of anaerobic capacity? Sports Med. 40, 285–302.

Pilarski, J.Q., Solomon, I.C., Kilgore Jr., D.L., Hempleman, S.C., 2009. Effects of aerobic and anaerobic metabolic inhibitors on avian intrapulmonary chemoreceptors. Am. J. Physiol. Regul. Integr. Comp. Physiol. 296, R1576–1584.

Pluim, B.M., Zwinderman, A.H., van der Laarse, A., van der Wall, E.E., 2000. The athlete's heart. A meta-analysis of cardiac structure and function. Circulation 101, 336–344.

Richter, E.A., Kiens, B., Saltin, B., Christensen, N.J., Savard, G., 1988. Skeletal muscle glucose uptake during dynamic exercise in humans: role of muscle mass. Am. J. Phys. 254, E555–561.

Robach, P., Bonne, T., Fluck, D., Burgi, S., Toigo, M., Jacobs, R.A., Lundby, C., 2014. Hypoxic training: effect on mitochondrial function and aerobic performance in hypoxia. Med. Sci. Sports Exerc. 46, 1936–1945.

Sloth, M., Sloth, D., Overgaard, K., Dalgas, U., 2013. Effects of sprint interval training on VO_{2max} and aerobic exercise performance: a systematic review and meta-analysis. Scand. J. Med. Sci. Sports 23, e341–352.

Tadaishi, M., Miura, S., Kai, Y., Kawasaki, E., Koshinaka, K., Kawanaka, K., Nagata, J., Oishi, Y., Ezaki, O., 2011. Effect of exercise intensity and AICAR on isoform-specific expressions of murine skeletal muscle PGC-1alpha mRNA: a role of beta(2)-adrenergic receptor activation. Am. J. Phys. Endocrinol. Metab. 300, E341–349.

Tanaka, H., Seals, D.R., 2008. Endurance exercise performance in masters athletes: age-associated changes and underlying physiological mechanisms. J. Physiol. 586, 55–63.

Timmons, J.A., 2011. Variability in training-induced skeletal muscle adaptation. J. Appl. Physiol. 110, 846–853.

Turcotte, L.P., 2000. Muscle fatty acid uptake during exercise: possible mechanisms. Exerc. Sport Sci. Rev. 28, 4–9.

Weston, M., Taylor, K.L., Batterham, A.M., Hopkins, W.G., 2014. Effects of low-volume high-intensity interval training (HIT) on fitness in adults: a meta-analysis of controlled and non-controlled trials. Sports Med. 44, 1005–1017.

Yoshida, T., 1984. Effect of dietary modifications on lactate threshold and onset of blood lactate accumulation during incremental exercise. Eur. J. Appl. Physiol. Occup. Physiol. 53, 200–205.

CHAPTER 6

Speed as a Complex Conditional Ability

Speed is a conditional ability in addition to strength and endurance. However, it is strongly associated with explosive strength and coordination abilities, thus it is a complex ability. Factors that determine speed involve genetics, the nervous system, and the skeletal muscle system. It is worth noting that speed is relative, and its value needs to be interpreted in regard to a particular sport category since speed training for marathon athletes differs from that for sprinters. Both marathon athletes and sprinters need speed training since the fastest wins. External factors may also influence performance since a tennis game is slower on a clay court and a rally takes longer; players also have higher heart rates and lactic acid levels compared to when playing hard court games (Martin et al., 2011).

According to Newton's law the higher the forces that act on an object, the higher its acceleration is; thus one of the main goals of speed training is the development of explosive strength. Hence, factors that determine the force the muscle can generate all influence velocity, especially the amount of fast-twitch muscle fibers. Coordination and technique of a movement and characteristics of muscle contraction and relaxation all influence the velocity (Fig. 6.1). Characteristics of speed manifest differently in different motions.

6.1 FORMS OF SPEED

Speed as a complex ability is comprised of several elements, which rarely appear alone. Some of them are measurable and definable such as reaction time; others are so complex that they are difficult to describe, such as decision-making speed, or learning speed.

6.1.1 Reaction Time

Reaction time is a simple form of speed, and depends mainly on the nervous system. It is the time interval between a signal and the reaction to it—for instance, when the starter pistol is fired at the start of the 100 m.

The Physiology of Physical Training
https://doi.org/10.1016/B978-0-12-815137-2.00006-1
 111

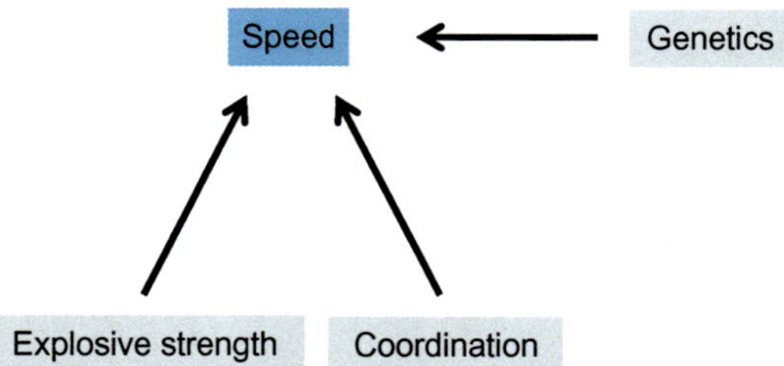

Fig. 6.1 Speed determining factors. Speed is greatly determined by genetics, dynamic force, and the quality of technical implementation of the motion.

The signal is perceived by the sensory system and the reaction evolves in the brain, then runs through the spinal cord to the muscles, resulting in contraction. Reaction time to sounds and visual information is on average 0.13–0.18 s, without consideration of speed of sound. It is determined by genetic factors and age, and it changes during effort; for instance, its value decreases/improves during loading and it is impaired by fatigue. Simple reaction time (reaction to a certain stimulus) of an average individual is 0.16–0.2. It can be improved by training; however, even the best sprinters cannot go below 0.1 s. The 100 m sprint is one of the shortest events, and the winner does not necessarily have the fastest reaction time.

Simple reaction time is the reaction time to a certain stimulus, whereas complex reaction time is the reaction to the right stimulus selected from many stimuli, and it increases reaction time.

6.1.2 Movement Speed

Movement speed in sport refers to the maximal speed of a nonlocomotory action such as throwing a ball, making a fencing strike, or serving in tennis. The move is started deliberately, and the goal is to reach the maximal speed (of a take-off, a shot). There are many technically simple to complex moves in sport that need to be executed as fast as possible be. The more complex a move, the more the accuracy of technical execution determines the speed. Since technical accuracy can be learned and dynamic strength can be improved, movement speed can be improved to some extent. It is also influenced by the equipment; for instance, the flexibility of a pole-vault athlete's pole influences the height of the jump, while the weight and string tightness of a tennis racket influences the speed at which a tennis ball can leave the racket. Movement speed does not comprise cyclic movements since they are associated with locomotion. Athletes with high movement speed do not necessarily have this in cyclic movements.

6.1.3 Acceleration Ability

Acceleration ability shows the rate of change of velocity of an athlete in a time interval or in a definite distance, thus starting from rest how fast they reach their maximal or submaximal speed. It is a very important ability for sprinters and in all ball games. The first couple of steps have significant importance in running (hence sprinting and ball games). Athletes with good acceleration ability touch the ground less on the first 5–10 m, and they use a higher force on the ground vertically and horizontally (Lockie et al., 2011). This emphasizes the importance of speed strength in this ability. Thus, sports in which acceleration is a key factor need to focus on the development of explosive strength. During acceleration from the start, for instance, in kayak events, athletes with good acceleration ability act high and quickly increase the force on the paddles, as measured by stamp force gauges (Fig. 6.2). The force acceleration slope diagram clearly predicts acceleration ability. Hence, explosive strength is the focus in acceleration training.

6.1.4 Maximal Locomotory Speed

This ability shows the speed of an athlete's cyclic locomotion, which is a characteristic of certain sports such as swimming, running, cycling, and rowing. Success is determined by maximal locomotory speed in shorter sport events such as 60–100 running, 50 m swimming, and 200 m kayak events. Technical execution is important in addition to excellent dynamic, explosive strength. Considering the abovementioned kayak acceleration example, the accelerated and moving kayak need slower and high force acted on the paddles with a greater area under the curve rather than quickly

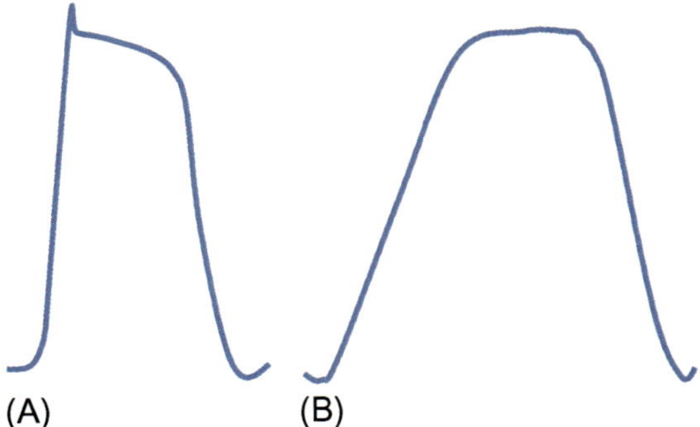

Fig. 6.2 Force acceleration slope of kayak athletes. Acceleration force on the paddles of kayak athletes with good acceleration ability (A) and average acceleration ability (B). The angle of the slope, the maximal force, and the area under the curve all have importance.

increasing and maximal force. The force placed on the blade of the paddle is determined by the position of the kayak (moving or static) and also the muscle fibers of the athlete. Fortunately, the author's country Hungary is very successful in kayak events; in kayak four athletes divide responsibility for starting, accelerating, and finishing the race, and for maintaining the boat's velocity during the race. Locomotory speed is determined by the technical execution of a movement, its effectiveness, and it is also influenced by the sport equipment, and external factors such as ski waxing.

6.1.5 Braking Ability

Braking ability is important in ball games during change of direction, and dribbling. Slowing high-velocity movements needs significant muscle force, which mostly involves eccentric muscle contraction. Considering the slowing of flexion motion in javelin-throwing, it is unintentional, a reflex to protect joints against injuries, whereas a change of direction or dribbling during ball games is intentional.

6.1.6 Decision Speed

In tactical sports such as ball games and martial arts, athletes are constantly in situations that require decision making since they need to choose the best action from their technical repertoire at every moment. This ability is mostly determined by the nervous system, especially the hippocampus. The often-mentioned game intelligence arises as the sum of the right decisions during a game. During the game, athletes need to choose the best alternative from the possible alternatives—for instance, to whom to pass the ball, or when to shoot for goals. The longer the decision making takes, the longer the opponent has to react. Fewer alternatives speed up decision making. This led to the development of a few routine tactics, where athletes in different positions know the next moves. In this case, choices are limited to a few alternatives, thus opponents have less time to react compared to a spontaneous game.

6.1.7 Learning Speed

Learning techniques successfully and applying them shows individual differences. Learning speed is determined by age, and techniques with different complexity have optimal ages for learning. The corpus callosum, a white flat bundle of neural fibers connecting the two hemispheres of the brain, is a key component of motor learning (Takeuchi et al., 2012). Visuomotoric response develops during the age of 4–12 months in infants

(Braddick and Atkinson, 2007), and is a basic element of effective motor learning. It is important to introduce many different movements to a child at the age of 2–3 years, because this promotes brain development. It has been demonstrated that fine motor skills of 3–4-year-old children predict mathematical and reading skills (Grissmer et al., 2010). Motor learning needs to be facilitated at an early age, which provides a strong base for efficient motor learning further on.

6.2 EXPLOSIVE TRAINING

Speed is a relative ability, and explosive training of marathon athletes differs from that of sprinters. Recruitment of fast-twitch muscle fibers is key to reach maximal or submaximal speed. These muscle fibers have higher activation thresholds compared to slow-twitch muscle fibers, thus activation require special situations such as maximal resistance or maximal speed. In the case of sport categories that use equipment, such as athletic throwing, handball, or water polo, maximal speed refers to the speed of motion with the equipment, e.g., a ball. Changing resistance gives the opportunity to recruit fast-twitch muscle fibers, for instance, the use of light or heavy weights, pulling or hauling during running, swimming with a kickboard. This results in higher speed due to adaptation. In addition to the high activation threshold of fast-twitch muscle fibers, they are fatigable, which limits the distance or sets where motions can be executed at maximal speed. Thus, development of maximal speed requires recruitment of fast-twitch muscle fibers in addition to exercises that improve technical execution, which influences internal resistance.

At maximal speed, anaerobic alactic energy sources such as CP and ATP provide energy, which regenerates in minutes. It is important to note that these exercises need to be scheduled when the athletes are in good condition, since it is necessary to recruit fast-twitch muscle fibers.

Speed is a complex ability since movement speed is influenced by the technical execution of the move. Thus, speed is a conditional ability that is influenced by coordination ability. Technical execution of a move can always be improved, since no perfect movement is perfect enough. Al Oerter mentioned in an interview that he had technical problems, thus emphasizing the importance of technical preparation. Yet he was a four-time Olympic gold medalist discus thrower, and 11 years after his last Olympic gold, at the age of 43, he set another record of 69.46 m. He always paid attention to improve his techniques.

One possible way to improve speed is advancing technical execution of a movement, via formation of a dynamic stereotype. A dynamic stereotype also means that speed can be automatically be set in addition to automatic reflex motions regulated by lower-level nervous structures. However, regaining an automatically set velocity can be challenging, a speed-limiting factor. Thus, setting and using many speed zones during training is important, including faster moves that will not be required in competitions. Supramaximal speed can be attained by aids or hauling, or during higher speeds such as downhill running.

Anaerobic interval and mini interval trainings (high-intensity interval training, HIT) is a popular training method to improve speed. This training method is discussed in detail in Chapter 5. It is important to note that these exercises can be done twice a day since CP and ATP energy resources can be recovered quickly. HIT exercise has beneficial effects on dynamic strength, as reflected by improved jump height (Buchheit and Laursen, 2013).

6.3 "R" EXTRA

It has been emphasized that the nervous system has a key role in speed ability. Theta and spike cells in the dorsal part of the hippocampus and fascia dentate influence running speed as demonstrated in animal experiments (McNaughton et al., 1983), and running speed influences the activation threshold of theta cells (Buzsaki, 2005). The hippocampus is important in learning, and hippocampal gamma oscillation is associated with speed in decision making (Montgomery and Buzsaki, 2007) and is also correlated with running speed. Higher speed is associated with higher gamma oscillation in CA1 pyramid cells, whereas at lower speed, activation of CA3 is more significant (Ahmed and Mehta, 2012). It is possible that elevating speed activates CA3 cells, and beyond a certain level, CA1 cells are recruited and their activity show linear elevation with increasing speed (Diba and Buzsaki, 2008).

Global positioning systems (GPS) have been used for many years in sports to track distance and speed. A review summarizing 30 articles published on this topic (Cummins et al., 2013) revealed information about performance of ball game teams. They compiled data and created groups based on speed between 0 and 36 km/h, and gathered information about running distance and speed of football/ice-hockey/rugby football players at certain positions, about acceleration and deceleration of players, and about the amount of clashes. Intensity and amplitude information from competitions is essential to select optimal training methods. The abovementioned study

Table 6.1 Reaction time (ms) and 20, 40, 60, and 80 m time (s) during 100 m dash

	Reaction time	20 m	40 m	60 m	80 m	100 m
Bolt (JAM)	0.146	2.89	4.64	6.31	7.92	9.58
Gay (USA)	0.144	2.92	4.70	6.39	8.02	9.71
Powell (JAM)	0.134	2.91	4.71	6.42	8.10	9.84
Bailey (ANT)	0.129	2.92	4.73	6.48	8.18	9.93
Thompson (TRI)	0.119	2.90	4.71	6.45	8.17	9.93
Chambers (UK)	0.123	2.93	4.75	6.50	8.22	10.00
Burns (TRI)	0.165	2.94	4.76	6.52	8.24	10.00
Patton (USA)	0.149	2.96	4.85	6.65	8.42	10.34

Table 6.2 Speed (m/s) during 100 m dash

	Start	20 m	40 m	60 m	80 m	100 m
Bolt (JAM)	0	6.92	11.43	11.98	12.42	12.05
Gay (USA)	0	6.85	11.24	11.83	12.27	11.83
Powell (JAM)	0	6.87	11.11	11.70	11.90	11.49
Bailey (ANT)	0	6.85	11.05	11.43	11.76	11.43
Thompson (TRI)	0	6.90	11.05	11.49	11.63	11.36
Chambers (UK)	0	6.83	10.99	11.43	11.63	11.24
Burns (TRI)	0	6.80	10.99	11.36	11.63	11.36
Patton (USA)	0	6.76	10.58	11.11	11.30	10.42
Mean value	0	6.85	11.05	11.54	11.82	11.40

provides detailed information about ball games promoting successful performance and selecting optimal speed training.

The 100 m run is a typical racing event. In Table 6.1, scores at the World Championships in Athletics in Berlin in 2009 are shown. Reaction times under normal circumstances do not influence the final outcome. All the athletes reached the highest speed at 80 m; the best sprinters need this distance to reach maximal speed (Table 6.2). Bolt had the highest acceleration, leading the field at 20 m.

6.4 SUMMARY

Speed is a complex ability, which is determined by technical knowledge and dynamic forces. Thus, speed training needs to include development of technical execution and dynamic strength. Speed comprises several elements, which are slightly independent and can be improved separately. These exercises need to be scheduled when the athletes are in good condition, because they are influenced strongly by the nervous system.

TEST QUESTIONS

(1) What factors determine speed?

(2) What exercises can improve speed?

(3) What elements is speed comprised of?

(4) What are the main characteristics of fast-twitch muscle fibers?

(5) What is a speed limiting factor?

BIBLIOGRAPHY

Ahmed, O.J., Mehta, M.R., 2012. Running speed alters the frequency of hippocampal gamma oscillations. J. Neurosci. 32, 7373–7383.

Braddick, O., Atkinson, J., 2007. Development of brain mechanisms for visual global processing and object segmentation. Prog. Brain Res. 164, 151–168.

Buchheit, M., Laursen, P.B., 2013. High-intensity interval training, solutions to the programming puzzle. Part II: anaerobic energy, neuromuscular load and practical applications. Sports Med. 43, 927–954.

Buzsaki, G., 2005. Theta rhythm of navigation: link between path integration and landmark navigation, episodic and semantic memory. Hippocampus 15, 827–840.

Cummins, C., Orr, R., O'Connor, H., West, C., 2013. Global positioning systems (GPS) and microtechnology sensors in team sports: a systematic review. Sports Med. 43, 1025–1042.

Diba, K., Buzsaki, G., 2008. Hippocampal network dynamics constrain the time lag between pyramidal cells across modified environments. J. Neurosci. 28, 13448–13456.

Grissmer, D., Grimm, K.J., Aiyer, S.M., Murrah, W.M., Steele, J.S., 2010. Fine motor skills and early comprehension of the world: two new school readiness indicators. Dev. Psychol. 46, 1008–1017.

Lockie, R.G., Murphy, A.J., Knight, T.J., Janse de Jonge, X.A., 2011. Factors that differentiate acceleration ability in field sport athletes. J. Strength Cond. Res. 25, 2704–2714.

Martin, C., Thevenet, D., Zouhal, H., Mornet, Y., Deles, R., Crestel, T., Ben Abderrahman, A., Prioux, J., 2011. Effects of playing surface (hard and clay courts) on heart rate and blood lactate during tennis matches played by high-level players. J. Strength Cond. Res. 25, 163–170.

McNaughton, B.L., Barnes, C.A., O'Keefe, J., 1983. The contributions of position, direction, and velocity to single unit activity in the hippocampus of freely-moving rats. Exp. Brain Res. 52, 41–49.

Montgomery, S.M., Buzsaki, G., 2007. Gamma oscillations dynamically couple hippocampal CA3 and CA1 regions during memory task performance. Proc. Natl. Acad. Sci. U. S. A. 104, 14495–14500.

Takeuchi, N., Oouchida, Y., Izumi, S., 2012. Motor control and neural plasticity through interhemispheric interactions. Neural Plast. 2012, 823285.

CHAPTER 7

Fundamentals of Joint Flexibility

7.1 JOINT FLEXIBILITY

Force is generated by muscles, and it is transmitted to the bones through the tendons resulting in joint movements and locomotion. Amplitude of joint motion influences acceleration over a long distance, or it makes the movement more esthetic. It has three critical factors: skeletal muscle, connective tissue, and joint type. Muscles have high stretchiness, unlikely connective tissue including tendons. Tendons are less flexible and have a lower metabolic rate and moderate blood supply, thus their adaptation to loading is much slower than muscles. According to Hill's three-element muscle model, tendons are passive elastic elements, and so they are stretched during dynamic efforts.

Generally speaking, greater amplitude of movements positively influences physical performance, thus flexibility exercises are an important part of physical training and warm-up. A method to increase muscle flexibility is stretching, which has different outcomes if applied before or after loading. After submaximal aerobic training it improves the amplitude of movements (Yamaguchi et al., 2007), and reduces the incidence of injuries (McHugh and Cosgrave, 2010), whereas before a race it may result in injury (Weldon and Hill, 2003). We shall now look at the factors that determine flexibility of the muscles and tendons.

7.2 STRETCHINESS OF MUSCLES AND TENDONS

Skeletal muscle and associated connective tissue, e.g., tendons, are comprised of flexible, elastic, and less elastic elements. Myofibrils and elastin are stretchable, while elements that are abundant in collagen, e.g., tendons, membranes, and fascia, are more rigid. Joint flexibility determines sport performance, thus stretching is vital during warm-up and to promote better performance. Stretching has two main effects: neuronal and viscoelastic. The neuronal effect is exerted through myotatic reflex (stretch reflex), while the viscoelastic effect is mechanical in nature and it is exerted via the flexible, elastic, and rigid elements. The main component of connective tissue is collagen, which has several types. It is high-tensile; however, it is moderately adaptable compared to any other structural components of muscle. Elasticity of collagen is determined by many factors such as age, temperature, and cross-linkage of proteins; the latter increases upon training. Lower liquid content inside muscle fibers also impairs the stretchiness of collagen.

The duration of stretching is important; 6–15 s is needed for total relaxation of a muscle, thus stretching needs not to exceed 20 s. In addition, velocity is important since quickly executed stretching increases muscle tension, thus stretching needs to be executed slowly to reach a high amplitude. After stretching, muscles and tendons keep their stretched length for a while due to viscous elements; the original length is recovered by flexible and elastic elements with time. However, due to frequent stretching, the amplitude of movements increases.

Acute, slow, passive stretching results in decreased neuromotor activity and altered myotatic reflex leading to higher stretchiness, and also influences joint flexibility. The muscle spindle detects the length of the muscle (Fig. 7.1).

Quick stretches stimulate muscle contraction and impair stretchiness. Muscle spindles are also responsible for reciprocal inhibition, which results in relaxation of antagonist muscles during agonist contraction. For instance, contraction of the biceps femoris results in relaxation of the triceps femoris. The tension signal from the muscle spindle is also transmitted to the brain (Fig. 7.2).

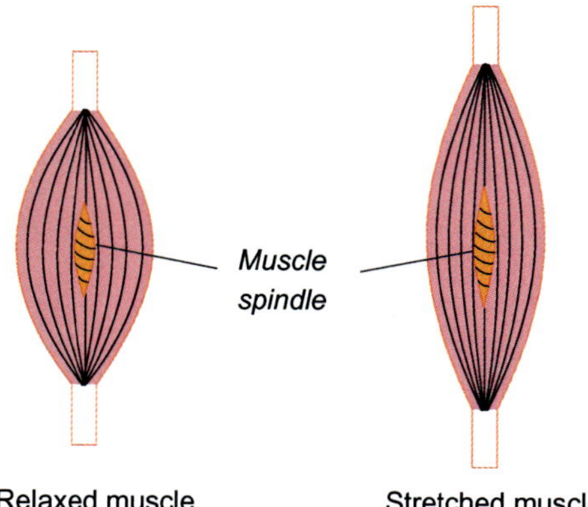

Fig. 7.1 Location of the muscle spindle. The muscle spindle detects the tension in the muscle.

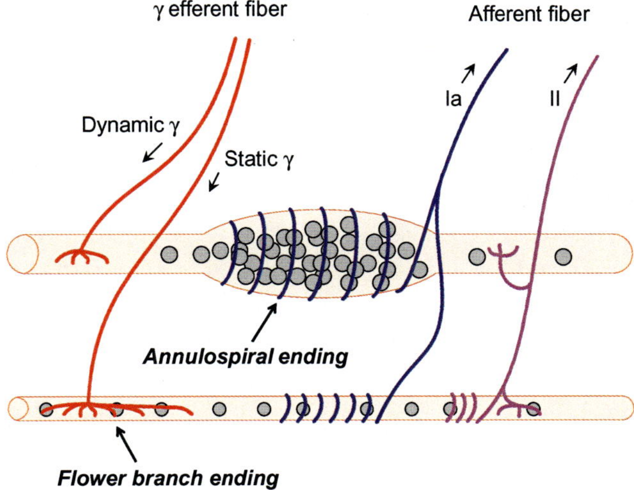

Fig. 7.2 Structure of the muscle spindle.

The Golgi tendon organ detects changes in tension where the muscle and the tendon are joined. During contraction, the tension generated by the muscle is transmitted to the tendons; this is very important since it signals the muscle to relax (Fig. 7.3).

Afferent fibers originated in the Golgi tendon organ transmit signals to the spinal cord and activate a reflex that regulates muscle tension. Thus, tension in the tendon regulates the tension of the muscle through the efferent fibers. The Golgi tendon organ is activated by stretching over 6 s, or

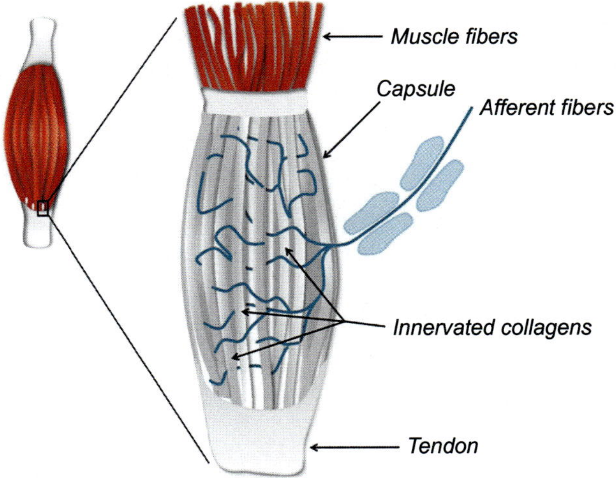

Fig. 7.3 Structure of the Golgi tendon organ.

by intense stretch or intense muscle contraction. Types and characteristics of muscle tension (slow, quick, accelerating, or decelerating) are also important factors since they directly influence stretchiness via the myotatic reflex.

7.3 PROPRIOCEPTIVE NEUROMUSCULAR FACILITATION

Static stretching utilizes gravitation or external force. Muscles and connective tissue (e.g., tendons) respond well to this type of passive stretching, and there is low risk of injury. Muscles can also relax during passive stretching, which broadens the limits of stretchiness.

Dynamic stretching induces different responses of the muscle-tendon complex. It utilizes the force generated by the antagonist muscles. Contraction of antagonists can be quick, e.g., kicking and thigh flexors stretch. This type of stretching occurs during actual movements, and stretching is limited due to the stretch reflex, which increases the tension of the stretched muscle.

Proprioceptive neuromuscular facilitation (PNF) is a method that switches off the stretch reflex, and so the muscle-tendon complex can be stretched in this relaxed state. The Golgi tendon organ needs to send signals to the spinal cord that induce relaxation of the muscle. This increases stretchiness and decrease the risk of injury. There are several types of PNF method.

Reversal antagonist methods are techniques that allow for agonist contraction before antagonist contraction without relaxation.

Repeated contraction methods are done by repeated isotonic contractions from the lengthened range, induced by quick stretches and enhanced by resistance; performed through the range or part of range at a point of weakness.

Agonist reversal methods, uses repeated isotonic contractions from the lengthened range, induced by rapid stretches and enhanced by resistance; performed through the range or part of range at a point of weakness.

Contract relax method generally performed at a point of limited range of motion in the agonist pattern. Strong, small range isotonic contraction of the restricting muscles (antagonists) with emphasis on the rotators is followed by an isometric hold. The contraction is held for 6–8 s and is then followed by relaxation and movement into the new range of the agonist pattern.

During the contraction-relaxation method, muscle is passively stretched to a maximal degree, and then static resistance (e.g., counterforce by a partner, or weight) induces shortening of contractile elements, while muscle length is unchanged. Contraction needs to last over 6 s, after which the muscle relaxes and can be stretched. In contrast, passive stretching needs to last over 20 s. This type of stretching should ideally be repeated 3–4 times.

Isometric stretching is another type of stretching method in addition to PNF. Passive stretching to a maximal degree followed by static contraction over 6 s, then stretching lasting over 20 s, is essential to successful results. The amplitude of movement widens following each stretch.

Another PNF method, a variation of the previous exercise, comprises of antagonist muscle contraction of 6 s following static contraction, and this is followed by stretching of the agonist. In this case, reciprocal inhibition occurs; the agonist muscle relaxes during antagonist contraction. The 6-s duration is recommended in order to switch off the myotatic reflex during stretching and contraction.

7.4 ADDITIONAL LIMITING FACTORS OF JOINT FLEXIBILITY

Joint flexibility require stretchiness of skeletal muscle and connective tissue; however, the primary factor determining joint flexibility is the type of joint. Ball joints such as hip and shoulder joints allow the greatest range of motion, while hinge joints between the phalanges limit movement along only one axis. Joint flexibility is determined by age, sex, and temperature.

It has been thought that there is reverse correlation between strength of a muscle and flexibility. The opposite is true since active flexibility is strength dependent, thus muscle strength improves flexibility. The confusion is caused by observation of athletes preoccupied with strength development and neglecting flexibility training. Sports such as weight lifting, which requires both maximal strength and flexibility (primarily hip joint and shoulder joint flexibility), demonstrate that the two ability can be developed together.

Flexibility directly influences physical performance, thus flexibility training is important. Children have high flexibility, whereas the elderly have limited flexibility. Flexibility to a certain level reduces the risk of injuries; however, extreme flexibility may adversely affect performance and increase the risk of injuries (Allen, 2014). Outstanding tennis players, such as retired Kim Clijsters and still active Novak Djokovic, have high flexibility, which allows them to exert brilliant techniques that are impossible for other athletes to emulate.

7.5 "R" EXTRA

Six-seven strength training followed by stretching significantly improved the speed of serving of young tennis players, and decreased the risk of injuries compared to a group participating only in strength training (Fernandez-Fernandez et al., 2013). The combined training widened the range of movements, so they could accelerate the tennis racket over a longer distance and hit the ball with a higher velocity. Thus, increasing the range of movements has a positive effect in tennis according to those authors.

Regeneration with cold and hot water baths is a generally used method following training; however, measurements did not indicate increased stretchiness of the thigh flexors (Burke et al., 2001). Selection of optimal flexibility training depends on the type of sport, and this ability also needs to be monitored frequently.

7.6 SUMMARY

Joint flexibility is strongly correlated with performance in certain sports. Every sport requires some degree of joint flexibility, and some need a high degree of flexibility. During flexibility training, the myotatic (stretch) reflex needs to be considered since its switching off through different stretching methods is an important element in flexibility development.

TEST QUESTIONS

(1) What determines range of motion?
(2) What role does the Golgi tendon organ have?
(3) What is the physiological base of stretching?
(4) What is PNF?
(5) What is the correlation between flexibility and strength?
(6) Why is joint flexibility important?

BIBLIOGRAPHY

Allen, D.J., 2014. Treatment of distal iliotibial band syndrome in a long distance runner with gait re-training emphasizing step rate manipulation. Int. J. Sports Phys. Ther. 9, 222–231.

Behm, D.G., Chaouachi, A., 2011. A review of the acute effects of static and dynamic stretching on performance. Eur. J. Appl. Physiol. 111, 2633–2651.

Bien, D.P., 2011. Rationale and implementation of anterior cruciate ligament injury prevention warm-up programs in female athletes. J. Strength Cond. Res. 25, 271–285.

Burke, D.G., Holt, L.E., Rasmussen, R., MacKinnon, N.C., Vossen, J.F., Pelham, T.W., 2001. Effects of hot or cold water immersion and modified proprioceptive neuromuscular facilitation flexibility exercise on hamstring length. J. Athl. Train. 36, 16–19.

Fernandez-Fernandez, J., Ellenbecker, T., Sanz-Rivas, D., Ulbricht, A., Ferrautia, A., 2013. Effects of a 6-week junior tennis conditioning program on service velocity. J. Sports Sci. Med. 12, 232–239.

Goldman, E.F., Jones, D.E., 2010. Interventions for preventing hamstring injuries. Cochrane Database Syst. Rev. (1) CD006782. https://doi.org/10.1002/14651858.CD006782.pub2.

Jenkins, J., Beazell, J., 2010. Flexibility for runners. Clin. Sports Med. 29, 365–377.

Kallerud, H., Gleeson, N., 2013. Effects of stretching on performances involving stretch-shortening cycles. Sports Med. 43, 733–750.

Lauersen, J.B., Bertelsen, D.M., Andersen, L.B., 2014. The effectiveness of exercise interventions to prevent sports injuries: a systematic review and meta-analysis of randomised controlled trials. Br. J. Sports Med. 48, 871–877.

Liederbach, M., 2010. Perspectives on dance science rehabilitation understanding whole body mechanics and four key principles of motor control as a basis for healthy movement. J. Dance Med. Sci. 14, 114–124.

McHugh, M.P., Cosgrave, C.H., 2010. To stretch or not to stretch: the role of stretching in injury prevention and performance. Scand. J. Med. Sci. Sports 20, 169–181.

Weldon, S.M., Hill, R.H., 2003. The efficacy of stretching for prevention of exercise-related injury: a systematic review of the literature. Man. Ther. 8, 141–150.

Winters, M., Eskes, M., Weir, A., Moen, M.H., Backx, F.J., Bakker, E.W., 2013. Treatment of medial tibial stress syndrome: a systematic review. Sports Med. 43, 1315–1333.

Yamaguchi, T., Ishii, K., Yamanaka, M., Yasuda, K., 2007. Acute effects of dynamic stretching exercise on power output during concentric dynamic constant external resistance leg extension. J. Strength Cond. Res. 21, 1238–1244.

CHAPTER 8

Diet and Sport

8.1 DIET AND METABOLISM

Muscle contractions consume energy, which is provided by carbohydrates, lipids, and rarely proteins. High intensity training demands large amounts of blood sugar and glycogen from muscles to produce ATP, whereas low intensity training consumes free fatty acids (Romijn et al., 1993). The level of fatty acids increases in the blood during rest after high intensity training, which demands carbohydrates. Thus, consumption of fatty acids is elevated during recovery (Kimber et al., 2013). This example implies that intensity and duration of exercises and recovery intervals influence metabolism in a complex way, and selecting the right diet can be challenging. Diet has special importance when preparing for a competition or regeneration, and influences sporting performance (Fig. 8.1). This chapter focuses on diet design in order to maximize performance.

Vitamins and minerals are also important elements of a healthy diet in addition to carbohydrate, fat, and protein intake. Activity of enzymes involved in metabolism is a key factor in metabolic processes, which are comprised of anabolic ("building up") and catabolic ("breaking down") processes. Anabolic processes demand energy to build up and renew cellular organelles, whereas catabolic processes produce chemical energy from breaking down organic molecules. Side products of breakdown such as lactic acid, ammonia, ADP, and Pi can directly influence physical performance.

The daily energy demand of an average 20-year-old male (70 kg) is 2000–2400 kcal, while a female's is 1800–2200 kcal. Most of the energy is consumed by basic metabolic processes, maintenance of body temperature, and optimal functioning of organs, which show significant individual differences (Mifflin et al., 1990). A lower caloric intake decreases the amount of energy consumed by basic metabolic processes, and thus a lower basic metabolic rate. The energy demands of physical exercising cover a wide range; for instance, the energy demands of a marathon runner and a sprinter differ hugely. Type of fatigue strongly correlates with different types of catabolic processes and side products (e.g., lactic acid, ammonia), thus we distinguish

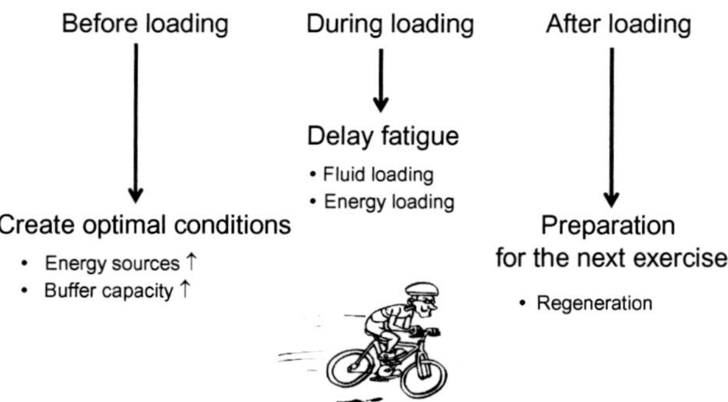

Fig. 8.1 Sport diet intervention focus. A sport nutrition allows for a wide range of intervention. Before training, energy resources need to be refilled and optimized, and increase buffer capacity. During training, fatigue needs to be delayed with energy and liquid intake. Following training, the focus should be on regeneration and preparation for the next exercise session.

sports in which aerobic endurance dominates from sports in which strength and speed are the focus.

8.2 DIET AND AEROBIC ENDURANCE

In sports in which aerobic endurance is in the focus, the main causes of fatigue are the depletion of muscle glycogen stores, decreased blood sugar level (hypoglycemia), and dehydration due to long-lasting low–intensity ($VO_2max < 60\%$) exertion. Significant loss of liquid volume (2%–3% of total body weight) impairs physical performance significantly (Sawka and Noakes, 2007) (Fig. 8.2).

Thus, sufficient liquid intake is essential during training or a competition. Insufficient liquid intake may result in hyperthermia (Fig. 8.3).

Liquid intake of 5–7 mL/kg of body weight 4 h before competitions and long training sessions can optimize the body liquid content. Further liquid intake is recommended if there is no urination in 2 h or its color is dark, indicating a concentrated quality (Sawka and Noakes, 2007). Hyperhydration may be beneficial in the case of a warm day or high humidity; however, it may cause vomiting, resulting in loss of sodium and impaired performance.

Liquid and nutrition intake during competition and long-lasting training are recommended. Liquid carbohydrate supplement (15, 30, and 60 g glucose) improves performance, and the improvement rate correlates with the amount of consumed carbohydrate (Smith et al., 2010). When selecting supplements, one should consider the combination of carbohydrates with

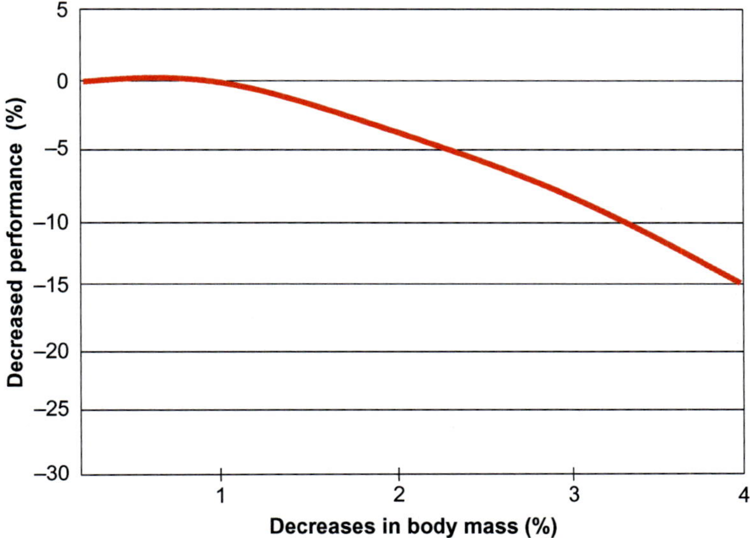

Fig. 8.2 Effect of weight loss (due to dehydration) on physical performance. Lack or insufficient liquid intake during long-lasting loading impairs physical performance.

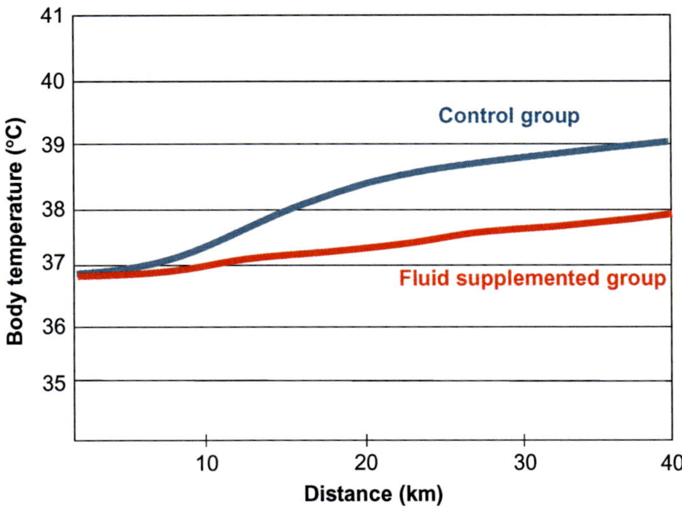

Fig. 8.3 Elevation of body temperature during physical training. The graph shows the body temperature changes of two groups in a 40 km bicycle training; the *blue line* marks the body temperature of athletes without liquid intake, while the *red line* marks liquid intake. *(Modified from Berkulo, M.A., Bol, S., Levels, K., Lamberts, R.P., Daanen, H.A., Noakes, T.D., 2015. Ad-libitum drinking and performance during a 40-km cycling time trial in the heat. Eur. J. Sport Sci., 1–8.)*

different absorption time that is the most beneficial. These supplements increase blood sugar level, which delays the development of fatigue of the nervous system during long-lasting training or competitions.

Long-lasting loading depletes glycogen stores of muscle, and recovery takes several days, thus food intake following training has special importance since it influences regeneration and optimal preparation.

Caffeine intake before a competition may be beneficial; it reaches its peak in the blood 30–90 min, after intake, which needs to be considered when scheduling its consumption (Cox et al., 2002). It improves endurance (1–3 mg/kg) since it stimulates free fatty acid combustion during training (Tarnopolsky, 2010), so it spares glycogen stores, which can be used later in the finish. Caffeine also stimulates Ca-ion reuptake by the sarcoplasmic reticulum, thereby decreasing the incidence of injuries due to delayed muscle relaxation.

Recovery of muscle glycogen stores is essential during a football European or world competition and serial loading, for example, since it directly influences sport performance. It is challenging when not even 1 day of resting is available. Significant carbohydrate intake following training increases muscle glycogen synthesis (Casey et al., 2000), which is an important part of regeneration. Timing is extremely important in this case, since insulin sensitivity is high and it stimulates glycogen buildup from consumed carbohydrate (Goodyear et al., 1990). Carbohydrate intake >2h after training deteriorates glycogen buildup by 50%. Some types of carbohydrate also influence glycogen buildup following training. Carbohydrates with a high glycemic index are more beneficial immediately after training since they stimulate insulin production, and insulin is an essential element of glycogen synthesis (Burke et al., 1993). Glucose has a higher glycemic index than fructose; chocolate also has a high glycemic index. Carbohydrates with a low glycemic index are recommended for glycogen buildup 1 day before loading.

A carbohydrate and protein mix seems to be more beneficial than carbohydrate intake alone. Insulin sensitivity can be increased by a leucine-phenylalanine-tyrosine-carbohydrate mix (Floyd et al., 1966). Saunders et al. have shown that aerobic performance of bicycle athletes was better on the second day after consuming a carbohydrate-protein mix compared to athletes consuming only carbohydrate. Thus, glycogen buildup after training is more efficient with a carbohydrate-protein mix (Saunders et al., 2004). It shows the importance of diet in physical performance.

Fat also provides energy when the duration of the training exceeds 1 h. Trained athletes mobilize and combust fatty acids more efficiently.

The recommended daily fat intake is 2 g/kg body weight (Decombaz et al., 2000); anything over this amount adversely effects glycogen buildup.

8.3 DIET AND STRENGTH AND SPEED

In strength- and speed-dependent sports, the intensity of physical exertion is high compared to endurance sports, thus the primary sources of energy and the type of fatigue also differ. Therefore, dietary requirements of power athletes differ from those of endurance athletes. One goal of strength training is muscle cross-sectional area growth, which is not an aim of endurance training. Designing a diet for power athletes is a challenge compared to creating a diet for endurance athletes. Mustafa Ismail is a Guinness recorder bodybuilder with a biceps perimeter of 79 cm; he consumes 5000 kcal daily in a diet which naturally differs in composition from the diet of a Tour de France athlete, for example. Training intensity over VO_2max 75%–80% stimulates breakdown of carbohydrates almost exclusively. Lower intensity training also stimulates the combustion of free fatty acids in addition to muscle glycogen breakdown. It has been known since the late 1960s that high carbohydrate intake increases the glycogen content of muscles and delays the development of fatigue, thereby improving performance (Bergstrom et al., 1967). Thus, a diet with high carbohydrate content is recommended in sports that require glycogen and during preparations for competition, which comprise frequent and high amplitude exercising. Thus, a daily carbohydrate intake for males of 6–12 g/kg body weight (6 g/kg for females) is recommended (Stellingwerff, 2013). The diet of power athletes and bodybuilders needs to include a high amount of protein: 1.5–1.7 g/kg body weight (Tarnopolsky, 1999).

In speed sports, lactic acid production is significant, resulting in low pH of blood and muscle. Low pH contributes to muscle fatigue, which can be influenced by diet only to a low extent. Beta-alanine intake for weeks before competition/training prevents pH from decreasing (Baguet et al., 2010). The underlying cause behind this is an elevated carnosine level (dipeptide made up from beta-alanine and histidine), which plays a role in the buffer capacity of the muscle. To increase the carnosine level significantly, an athlete's diet needs to contain 3–6 g beta-alanine daily for 4–8 weeks (Baguet et al., 2010). Another possible way to balance pH is through the intake of sodium bicarbonate ($NaHCO_3$), which also increases buffer capacity. This is extracellular buffer capacity, whereas the beta-alanine effect is called intracellular buffer capacity.

Intense physical training promotes lactic acid production, and it is transferred into the bloodstream via active transport and conjunction with H^+, which strongly correlates with the development of fatigue. Sodium bicarbonate intake increases the buffer capacity of the blood, resulting in moderate pH decline. It needs to be consumed 1–6 min. Before loading (0.3 g/kg body weight/day) (McNaughton, 1992; McNaughton et al., 2008), and it results in improved performance in the 2–4 min. Immediately after intake. Extra short loading (10–30 s.) does not show any improvement. Taking beta-alanine and sodium bicarbonate jointly ensure a better buffer capacity, since they offer a broadened capacity, including both extracellular and intracellular (Sale et al., 2011).

Other options to improve performance in strength and speed sports include creatine monohydrate intake. Muscle contains significant amounts of creatine, 130 mmol/kg (dry muscle weight): which 30%–40% free and 60%–70% present in phosphorylated form (CP) (Sale et al., 2011). CP reacting with ADP provides ATP for a short time. Intake of creatine monohydrate (20 g/day) for 5 days results in a 15%–20% increase in muscle creatine content (Harris et al., 1992), which improves speed performance (Kreider, 2003).

In summary, diet significantly influences physical performance and provides an opportunity to increase performance legally. Trainers need to keep up with the newest discoveries in nutrition science and apply that knowledge.

8.4 DIET AND WEIGHT-CLASS SPORTS

Weight-class sports such as judo, wrestling, boxing, and weight-lifting require special attention in regard to diet. During preparation, athletes' body weight is likely to be over the weigh-in value, thus training also needs to focus on losing weight. Weight loss, however, adversely affects performance, which needs to be considered. Temporary weight loss includes dehydration; however, a deficit of total body water may have serious consequences. Methods include sauna, heat cabins, wearing plastic or rubber clothing during training, and sometimes voluntary vomiting. Weight loss in this way may reach 10% of total body weight, and a competition cycle with several competitions in a year challenges the body to an extreme degree. Quick weight loss has serious physiological consequences: it may result in impaired functions of the immune system, endocrine issues, decalcification, ion imbalances, depression, injuries, and nervous system dysfunctions. Weigh-in is scheduled shortly before a contest in weight-lifting, 3–6 h before a game in judo, and a day before in wrestling and boxing, according to recent rules. Therefore there is time for liquid intake; however, even this longer time is not enough for total regeneration. Thus, quick weight loss should be avoided and instead, gradual weight loss with continuous control is recommended during preparation; thereby only mild impairment is present during aerobic performance.

8.5 VITAMINS, MINERALS, AND SPORT PERFORMANCE

High-altitude conditions impair gas exchange due to lower atmospheric pressure. The human body adapts to such an environment by increasing the number of circulating red blood cells. This process demands an iron supply, since hemoglobin in the red blood cells contain iron. High-altitude conditions also influence anabolic and metabolic processes; thus designing an optimal diet is challenging, as it needs to emphasize carbohydrate and iron intake. Antioxidant intake is popular among athletes since intensive training increases the production of free radicals and oxidative damage. Vitamin C is water-soluble, while vitamin E is fat-soluble. Vitamin E is able to infiltrate cellular membranes and defend the cells against oxidants, while vitamin C exerts its antioxidant actions in cellular compartments. There is no experimental data indicating whether antioxidant intake improves performance or stimulates regeneration following loading. In some cases, antioxidants delayed the development of fatigue; however, this was accompanied by decreased strength. Antioxidants ameliorate or cease oxidative stress by neutralizing oxidants and reducing their toxic effects. Free radicals deteriorate enzyme activity, alter membrane fluidity, and damage DNA. Thus, high amounts of free radicals impair cellular functions, which may result in cell death. In addition to nonenzymatic antioxidants, cells are equipped with an enzymatic antioxidant system. The general goal of antioxidant intake is to reduce oxidative damage during training and improve cellular functions. It is important to note that a low level of oxidants plays an important role in cellular signaling and normal cellular functions. Thus, antioxidant intake may impair normal cellular functions under certain circumstances.

It has been argued that antioxidant intake is a double-edged sword, and that it may have positive or negative effects on physical performance. This effect is described well with hormesis; antioxidant intake may have positive or negative effects depending on the circumstances and timing (Fig. 8.4).

Thus, antioxidant intake is recommended in cases of suppressed vitaminosis. An optimal level of antioxidants shows individual differences, and

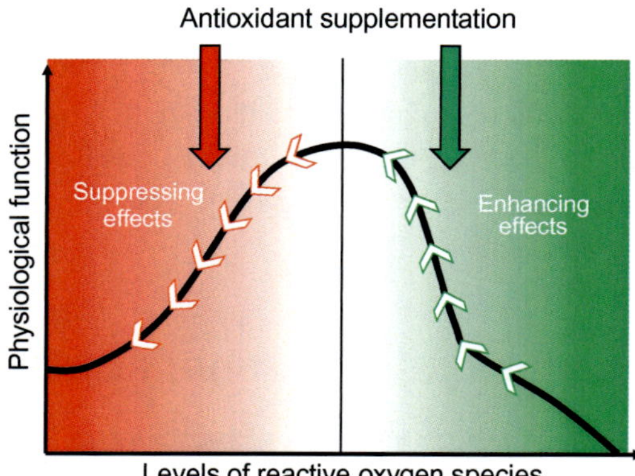

Fig. 8.4 Physiological effect of antioxidant intake in different phases. Free radicals are essential in muscle contraction and in other physiological functions. They facilitate muscle contraction to some extent; however, a high level of oxidants promotes fatigue. Thus, antioxidant intake improves or impairs physiological functions, and ensuring adequate antioxidant intake is challenging.

age and lifestyle also have an influence. Proving vitamin deficiency requires complex biochemical assays. To define optimal levels of vitamin intake individually is challenging; the recommended amount prevents avitaminosis. The optimal level of antioxidant regarding lifestyle and age is still elusive, and similarly so regarding physical performance. Our animal experiments have shown that resveratrol administration, which is a strong antioxidant extracted from red grapes, decreased the running performance of rats with low running capacity; however, rats with high running capacity showed improved physical performance by 20% (Hart et al., 2014, 2013). This clearly shows the effect of genetic differences on adaptation responses to diet.

Iron is essential in oxygen transport (hemoglobin and myoglobin contain iron) and metabolism (cytochrome enzymes of mitochondrial electron chain contain iron), and the daily recommended amount is 18–45 mg. Iron has a direct effect on sport performance due to its role in oxygen transport and metabolism. For instance, the same training intensity causes a higher heart rate in athletes with lower hemoglobin levels since their hearts contract more frequently to assure the delivery of the same amount of oxygen. Iron deficiency manifests itself in three steps. First, depleted iron sources decrease the ferritin level in the blood, which results in impaired physical performance if its value drops below $12\,\mu g/L$. In the second phase, transferrin concentration is increased and this influences the formation of red blood cells. The third step is a condition called anemia, indicated by a decreased amount of red blood cells in the circulation, and a hemoglobin

(Hb) concentration of below 12 mg/dL in females and 13 mg/dL in males (Beard and Tobin, 2000). A decreased amount of cytochrome enzymes in skeletal muscle influences metabolic processes and thereby impairs physical performance.

Frequent training increases the chances of developing iron deficiency due to enhanced hemolysis (red blood cell breakdown) and decreased iron absorption. Female athletes may have a higher chance of developing anemia due to menstrual loss and inadequate intake of iron. Iron lost through sweating is significant in the first 30 min of loading (Waller and Haymes, 1996). An optimal iron level is a key factor in both recreational and professional sports, and an individual's iron level should be monitored before iron intake.

Magnesium is another important mineral, which plays a role in carbohydrate, fat, and protein metabolism, and also in immune and endocrine functions. It influences cortisol level, which is a catabolic hormone affecting muscle mass. The daily recommended magnesium intake is 320 mg. Magnesium deficiency results in muscle and nervous system disturbances. Magnesium intake decreases cortisol levels, resulting in improved anabolic functions during training (Golf et al., 1984). Zinc also an important mineral; it regulates metabolic and immune processes, affect DNS and RNA synthesis, and also influences antioxidant enzymes since superoxide-dismutase contains zinc. The daily recommended zinc intake is 7–11 mg. Zinc deficiency impairs physical performance; however, its performance enhancer effect is elusive.

Vitamin D is important in the intestinal absorption of calcium, phosphor, iron, zinc, and magnesium; thus it plays a role in bone formation and regeneration, oxygen transfer, and antioxidant functions. The skin is able to produce high amounts of Vitamin D from provitamins following exposure to sunlight (UV B), and nutritional intake is also important in addition, especially for children and the elderly. Vitamin D is measured indirectly through an intermediate product, calcidiol or $25(OH)D_2$ produced by the liver, and its reference value is 30.0–74.0 ng/mL in the blood (data from Medline). Daily vitamin D demand is 3000–5000 IU (international unit) or more for sport athletes (Holick, 2005). Vitamin D deficiency increases the risk of bone fractures and injuries (Lappe et al., 2008). Beneficial effects of "sun therapy" have long been known; in addition to its role in vitamin D generation, sun therapy positively influences mental functioning and also improves muscle functions (Girgis et al., 2013). Athletes doing indoor sports need to pay attention to ensure optimal vitamin D intake.

Vitamin C or ascorbic acid was discovered by a Hungarian scientist, Albert Szent-Györgyi. The daily recommended amount is about 100 mg, though Szent-Györgyi consumed 9000 mg daily. Vitamin C is able to deliver electrons and reduce molecules, which makes it an important agent in metabolic processes. It does not enhance physical performance; however, it has an antioxidant effect. Vitamin E also has an antioxidant effect, and unlike vitamin C, it is fat-soluble. Its recommended daily intake is about 400 IU. It appears in membranes and there exerts its antioxidant effect against free radicals. Vitamins C and E seem to reduce microscopic injuries in skeletal muscle and therefore regeneration time between trainings (Nikolaidis et al., 2012).

8.6 "R" EXTRA

Diet was in the focus in athletes' lives even in ancient Olympic times. Pythagoras, in addition to his mathematical achievements in the 3rd century BCE, trained an athlete, Eurymenes, and he recommended the intake of hundreds figs, cheese, and flour. Dromeus, a successful runner, coupled meat intake with fruits and vegetables as an enhancer of physical performance. Later, intake of meat and eggs, which are abundant in protein, was emphasized by athletes participating in the Berlin Olympics in 1936. The dietary habits of several athletes were registered around this era. Athletes from the USA consumed 3500 kcal energy a day, Brazilian athletes 2300 kcal, and there were also significant differences in protein, carbohydrate, and fat intake (Grivetti and Applegate, 1997). This brief historical retrospection shows that the correlation between diet and physical performance has been known for a long time. Today it is essential to utilize that accumulated data on diet.

Diet has a significant influence on mental functions. High fat and carbohydrate intake adversely affect brain functions in experimental animals, which was compensated by physical training (Molteni et al., 2004). Omega 3 fatty acids have a beneficial effect on brain function (Gomez-Pinilla and Ying, 2010), and their inadequate intake results in depression and dysfunctions (Giles et al., 2013). Omega 3 and other fatty acids stimulate Leydig-cells in the testis; these cells are responsible for testosterone production, which promotes anabolic processes (Volek et al., 1997). This may explain the muscle growth that occurs following Omega 3 fatty acid intake.

A healthy diet is extremely important since it influences overall function of the body. Gut flora also significantly influence overall health, since microorganisms affect the functions of organs directly. High-fiber foods such as konjac plant reduce the incidence of colon cancer and maintain healthy gut flora (Chua et al., 2010).

Timing of nutritional intervention is crucial and aims to enhance performance, delay fatigue, and enhance regeneration. The type of nutritional intervention should depend on the intensity and the duration of exercise. Prolonged exercise with an intensity of VO_2max 60%–80% with duration longer than 60 min. Relies on carbohydrate metabolism and availability; therefore carbohydrate loading before and during exercise improves performance. Rapid ingestion of carbohydrates >1.2 g/kg/h for 4–6 h after heavy exercise loads rapidly stimulates regeneration of muscle glycogen, especially when 0.2–0.5 g/kg/h protein is added to carbohydrate (Kerksick et al., 2017). Taking proteins before and even during training, which aims to increase muscle hypertrophy, provides enhanced muscle protein synthesis, which can also be increased by 30–40 g of casein intake 30 min. Before sleeping (Kerksick et al., 2017).

8.7 SUMMARY

A well-designed diet during a training period provides an optimal level of nutrition available during the competition. During a race, liquid and nutrition intake delay the development of fatigue and improve performance. Furthermore, we can improve the buffer capacity of the body, which improves performance in endurance sports. A balanced diet before and after training promotes the regeneration process. In weight-class sports, a well-designed diet is extremely important with continuous control during preparation, and quick weight loss should be avoided.

TEST QUESTIONS

1. How does diet determine the glycogen content of muscle?
2. What nutrition facilitates the buildup of glycogen following training?
3. What effects does caffeine have on performance?
4. What nutrition is recommended for muscle growth?
5. What nutrition is recommended during aerobic endurance training?
6. What sports gain benefits from creatine monohydrate intake, and why?
7. What is the best method to control weight in weight-class sports?

BIBLIOGRAPHY

Baguet, A., Koppo, K., Pottier, A., Derave, W., 2010. Beta-alanine supplementation reduces acidosis but not oxygen uptake response during high-intensity cycling exercise. Eur. J. Appl. Physiol. 108, 495–503.
Beard, J., Tobin, B., 2000. Iron status and exercise. Am. J. Clin. Nutr. 72, 594S–597S.

Bergstrom, J., Hermansen, L., Hultman, E., Saltin, B., 1967. Diet, muscle glycogen and physical performance. Acta Physiol. Scand. 71, 140–150.

Berkulo, M.A., Bol, S., Levels, K., Lamberts, R.P., Daanen, H.A., Noakes, T.D., 2015. Ad-libitum drinking and performance during a 40-km cycling time trial in the heat. Eur. J. Sport Sci, 1–8.

Burke, L.M., Collier, G.R., Hargreaves, M., 1993. Muscle glycogen storage after prolonged exercise: effect of the glycemic index of carbohydrate feedings. J. Appl. Physiol. 75, 1019–1023.

Casey, A., Mann, R., Banister, K., Fox, J., Morris, P.G., Macdonald, I.A., Greenhaff, P.L., 2000. Effect of carbohydrate ingestion on glycogen resynthesis in human liver and skeletal muscle, measured by (13)C MRS. Am. J. Phys. Endocrinol. Metab. 278, E65–75.

Chua, M., Baldwin, T.C., Hocking, T.J., Chan, K., 2010. Traditional uses and potential health benefits of *Amorphophallus konjac* K. Koch ex N.E.Br. J. Ethnopharmacol. 128, 268–278.

Cox, G.R., Desbrow, B., Montgomery, P.G., Anderson, M.E., Bruce, C.R., Macrides, T.A., Martin, D.T., Moquin, A., Roberts, A., Hawley, J.A., Burke, L.M., 2002. Effect of different protocols of caffeine intake on metabolism and endurance performance. J. Appl. Physiol. 93, 990–999.

Decombaz, J., Fleith, M., Hoppeler, H., Kreis, R., Boesch, C., 2000. Effect of diet on the replenishment of intramyocellular lipids after exercise. Eur. J. Nutr. 39, 244–247.

Floyd Jr., J.C., Fajans, S.S., Conn, J.W., Knopf, R.F., Rull, J., 1966. Insulin secretion in response to protein ingestion. J. Clin. Invest. 45, 1479–1486.

Giles, G.E., Mahoney, C.R., Kanarek, R.B., 2013. Omega-3 fatty acids influence mood in healthy and depressed individuals. Nutr. Rev. 71, 727–741.

Girgis, C.M., Clifton-Bligh, R.J., Hamrick, M.W., Holick, M.F., Gunton, J.E., 2013. The roles of vitamin D in skeletal muscle: form, function, and metabolism. Endocr. Rev. 34, 33–83.

Golf, S.W., Happel, O., Graef, V., Seim, K.E., 1984. Plasma aldosterone, cortisol and electrolyte concentrations in physical exercise after magnesium supplementation. J. Clin. Chem. Clin. Biochem. Zeitschrift fur klinische Chemie und klinische Biochemie 22, 717–721.

Gomez-Pinilla, F., Ying, Z., 2010. Differential effects of exercise and dietary docosahexaenoic acid on molecular systems associated with control of allostasis in the hypothalamus and hippocampus. Neuroscience 168, 130–137.

Goodyear, L.J., King, P.A., Hirshman, M.F., Thompson, C.M., Horton, E.D., Horton, E.S., 1990. Contractile activity increases plasma membrane glucose transporters in absence of insulin. Am. J. Phys. 258, E667–672.

Grivetti, L.E., Applegate, E.A., 1997. From Olympia to Atlanta: a cultural-historical perspective on diet and athletic training. J. Nutr. 127, 860S–868S.

Harris, R.C., Soderlund, K., Hultman, E., 1992. Elevation of creatine in resting and exercised muscle of normal subjects by creatine supplementation. Clin. Sci. 83, 367–374.

Hart, N., Sarga, L., Csende, Z., Koltai, E., Koch, L.G., Britton, S.L., Davies, K.J., Kouretas, D., Wessner, B., Radak, Z., 2013. Resveratrol enhances exercise training responses in rats selectively bred for high running performance. Food Chem. Toxicol. 61, 53–59.

Hart, N., Sarga, L., Csende, Z., Koch, L.G., Britton, S.L., Davies, K.J., Radak, Z., 2014. Resveratrol attenuates exercise-induced adaptive responses in rats selectively bred for low running performance. Dose-Response 12, 57–71.

Holick, M.F., 2005. The vitamin D epidemic and its health consequences. J. Nutr. 135, 2739S–2748S.

Kerksick, C.M., Arent, S., Schoenfeld, B.J., Stout, J.R., Campbell, B., Wilborn, C.D., Taylor, L., Kalman, D., Smith-Ryan, A.E., Kreider, R.B., Willoughby, D., Arciero, P.J., VanDusseldorp, T.A., Ormsbee, M.J., Wildman, R., Greenwood, M., Ziegenfuss, T.N., Aragon, A.A., Antonio, J., 2017. International society of sports nutrition position stand: nutrient timing. J. Int. Soc. Sports Nutr. 14, 33.

Kimber, N.E., Cameron-Smith, D., McGee, S.L., Hargreaves, M., 2013. Skeletal muscle fat metabolism after exercise in humans: influence of fat availability. J. Appl. Physiol. 114, 1577–1585.

Kreider, R.B., 2003. Effects of creatine supplementation on performance and training adaptations. Mol. Cell. Biochem. 244, 89–94.

Lappe, J., Cullen, D., Haynatzki, G., Recker, R., Ahlf, R., Thompson, K., 2008. Calcium and vitamin D supplementation decreases incidence of stress fractures in female navy recruits. J. Bone Miner. Res. 23, 741–749.

McNaughton, L.R., 1992. Sodium bicarbonate ingestion and its effects on anaerobic exercise of various durations. J. Sports Sci. 10, 425–435.

McNaughton, L.R., Siegler, J., Midgley, A., 2008. Ergogenic effects of sodium bicarbonate. Curr. Sports Med. Rep. 7, 230–236.

Mifflin, M.D., St Jeor, S.T., Hill, L.A., Scott, B.J., Daugherty, S.A., Koh, Y.O., 1990. A new predictive equation for resting energy expenditure in healthy individuals. Am. J. Clin. Nutr. 51, 241–247.

Molteni, R., Wu, A., Vaynman, S., Ying, Z., Barnard, R.J., Gomez-Pinilla, F., 2004. Exercise reverses the harmful effects of consumption of a high-fat diet on synaptic and behavioral plasticity associated to the action of brain-derived neurotrophic factor. Neuroscience 123, 429–440.

Nikolaidis, M.G., Kerksick, C.M., Lamprecht, M., McAnulty, S.R., 2012. Does vitamin C and E supplementation impair the favorable adaptations of regular exercise? Oxidative Med. Cell. Longev. 2012, 707941. https://doi.org/10.1155/2012/707941.

Romijn, J.A., Coyle, E.F., Sidossis, L.S., Gastaldelli, A., Horowitz, J.F., Endert, E., Wolfe, R.R., 1993. Regulation of endogenous fat and carbohydrate metabolism in relation to exercise intensity and duration. Am. J. Phys. 265, E380–391.

Sale, C., Saunders, B., Hudson, S., Wise, J.A., Harris, R.C., Sunderland, C.D., 2011. Effect of beta-alanine plus sodium bicarbonate on high-intensity cycling capacity. Med. Sci. Sports Exerc. 43, 1972–1978.

Saunders, M.J., Kane, M.D., Todd, M.K., 2004. Effects of a carbohydrate-protein beverage on cycling endurance and muscle damage. Med. Sci. Sports Exerc. 36, 1233–1238.

Sawka, M.N., Noakes, T.D., 2007. Does dehydration impair exercise performance? Med. Sci. Sports Exerc. 39, 1209–1217.

Smith, J.W., Zachwieja, J.J., Peronnet, F., Passe, D.H., Massicotte, D., Lavoie, C., Pascoe, D.D., 2010. Fuel selection and cycling endurance performance with ingestion of [13C]glucose: evidence for a carbohydrate dose response. J. Appl. Physiol. 108, 1520–1529.

Stellingwerff, T., 2013. Contemporary nutrition approaches to optimize elite marathon performance. Int. J. Sports Physiol. Perform. 8, 573–578.

Tarnopolsky, M.A., 1999. Protein and physical performance. Curr. Opin. Clin. Nutr. Metab. Care 2, 533–537.

Tarnopolsky, M.A., 2010. Caffeine and creatine use in sport. Ann. Nutr. Metab. 57 (Suppl. 2), 1–8.

Volek, J.S., Kraemer, W.J., Bush, J.A., Incledon, T., Boetes, M., 1997. Testosterone and cortisol in relationship to dietary nutrients and resistance exercise. J. Appl. Physiol. 82, 49–54.

Waller, M.F., Haymes, E.M., 1996. The effects of heat and exercise on sweat iron loss. Med. Sci. Sports Exerc. 28, 197–203.

CHAPTER 9

Physical Training and Prevention

The human genome, which determines basic physiological processes, may not been altered in the last 10,000 years. Thus, it is not surprising that the recent form of the genome has not been adapted yet to modern lifestyles. In Paleolithic times, obtaining protein and abundant nutrients was challenging for hunter-gatherers. Excellent endurance of prehumans made them successful in hunting. This evolutionary hypothesis is based on a conception that prehumans were chasing the prey until it was exhausted due to excess heat and impaired perspiration (Bramble and Lieberman, 2004; Zimmer, 2004). Prey suffered from hyperthermia and stopped fighting due to enzyme dysfunctions resulting from higher temperatures (42°C); the hunter was then able to kill the exhausted prey with a strike. This effective hunting method allowed prehumans to obtain high-quality protein sources.

A sheep is able to keep running on a treadmill for only 30 min at VO_2max 70% and 24°C external temperature. When its body temperature reaches 41°C, it stops running and collapses. This experimental observation goes some way to proving the above hypothesis. Thermoregulation of the human body is more advanced compared to other mammals since animals eliminate excess heat mainly by panting, which is difficult during running. A high external temperature impairs effective heat elimination. It is interesting that busmen in Botswana and Hazda tribe in, Tanzania, who live as hunter-gatherers similar to prehuman lifestyle, schedule hunting during the hot hours during the day

The energy intake of ancient *Homo sapiens* could be around 206 kJ/kg/nap (49 kcal/kg/nap) due to significant physical activity, which is higher compared to recent 136 kJ/kg/nap (32 kcal/kg/nap) energy demand. The human genome adapted to significant physical exertion and different environments and lifestyles. Prehumans were not able to eat three times a day—they might have been lucky to eat three times a week. Thus, the development of such genes that enable the body to accumulate energy and promote only moderate physical activity ("laziness genes") were essential. It is important to emphasize that these genes that developed during prehistoric times have determined physiological processes in recent modern times.

The Physiology of Physical Training
https://doi.org/10.1016/B978-0-12-815137-2.00009-7
 141

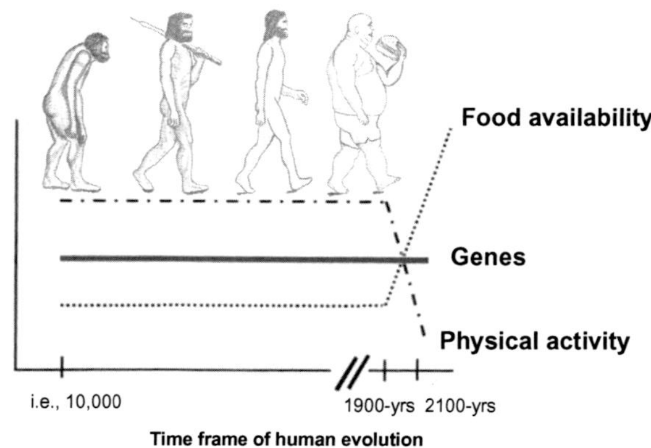

Fig. 9.1 Lifestyle changes and genes. Human genes have been altered only slightly following the evolvement of *Homo sapiens*. Physical activity, however, has been reduced significantly, and the availability of food has improved. These changes indicate that the relationship between genes and lifestyle has altered greatly in the last 100 years.

Daily intense physical activity, e.g., gathering food, migration, fighting, and hunting, has been replaced by physical inactivity in recent centuries (Fig. 9.1).

Modern, civilized lifestyles coupled with physical inactivity promote the development of lifestyle diseases. This chapter discusses the beneficial effects of frequent training on health and how it influences the incidence of lifestyle diseases; thus this chapter address prevention.

9.1 VO$_2$MAX AND CARDIOVASCULAR AND HEART DISEASES

The effect of physical training on cardiovascular and heart diseases has been widely explored. Cardiovascular dysfunctions are one of the leading causes of death in many countries including Hungary. There is no doubt that physical training prevents cardiovascular issues. Designing and selecting the frequency and quality of training is always a question of individual awareness; thus choosing a healthy lifestyle and preventing life-threatening diseases is an option available to everyone.

Experimental observations revealed the relationship between physical training and cardiovascular fitness, which is represented by VO$_2$max (Fig. 9.2).

An extensive meta-analysis including 1.3 million individuals revealed that there is a reverse correlation between cardiovascular fitness, VO$_2$max, and the incidence of cardiovascular diseases (Andersen, 1995; Blair et al.,

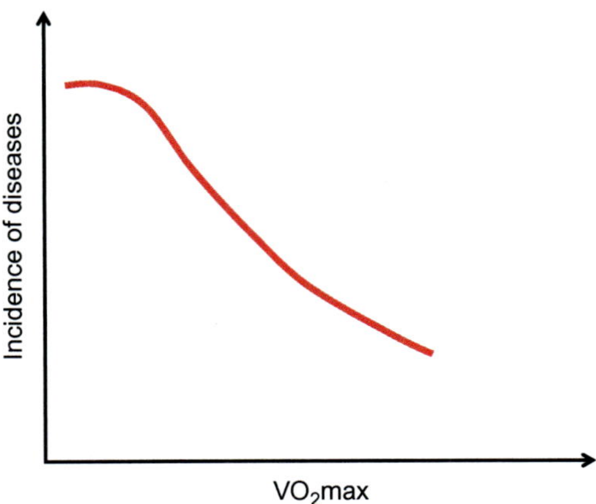

Fig. 9.2 Maximal oxygen uptake and incidence of diseases. There is a reverse correlation between cardiovascular fitness, VO$_2$max, and the incidence of cardiovascular diseases.

1995; Williams, 2001). Better physical fitness and higher VO$_2$max provide greater protection against cardiovascular disease. A 9-month program of lifestyle changes, comprised of frequent training and gradually increased training intensity, decreased blood pressure in individuals who suffered from metabolic syndrome with high blood pressure and obesity (Totsikas et al., 2011). In addition to better blood pressure, lipoprotein profile and insulin sensitivity also improved. A similar examination of 42–60-year-old males revealed that there is a strong correlation between VO$_2$max and resting heart rate, fasting blood insulin level, and the incidence of heart diseases (Laukkanen et al., 2009). Higher VO$_2$max was associated with a lower resting heart rate and lower fasting insulin level. The latter indicates that insulin sensitivity improved, which is a key point of diabetes prevention. High VO$_2$max also decreases the incidence of heart diseases (Laukkanen et al., 2009).

A Canadian longitudinal examination between 1981 and 1988 revealed that individuals who had a high VO$_2$max at the beginning of the study had lower incidence of obesity measured in 2002–2004 (Brien et al., 2007). It has been concluded that cardiovascular fitness predicts weight gain and risk of cardiovascular diseases. Cross-sectional examinations also revealed that cardiovascular fitness better predicts the lower incidence of risk factors (triglycerides, total cholesterol, waist circumference, blood pressure) of cardiovascular disease than physical activity. (It should be noted that data on physical activity were based on declaration by individuals, whereas VO$_2$max data were measured in a laboratory; Dvorak et al., 2000.)

In addition to lifestyle (frequent aerobic training in this case), genetic factors also influence the development of cardiovascular diseases. It has been demonstrated by studies, e.g., laboratory animals tested on a treadmill and divided into two groups depending on the distance they completed freely (Hussain et al., 2001). Offspring of the low capacity running (LCR) group showed higher blood pressure, more rigid blood vessels, and lower activity of heart mitochondrial aerobic enzymes compared to the high capacity running (HCR) group (Wisloff et al., 2005). LCR offspring were overweight, had insulin resistance, and had shorter lifespans (Koch et al., 2011, 2012). This investigation also showed higher VO_2max in the HRC group, which influences the development of cardiovascular diseases. Frequent training (1 h daily, 5 times a week, for 9 weeks) with VO_2max 70% intensity improved VO_2max in LCR group to the level of the HRC group; thereby the risk due to genetic predisposition was ameliorated (Hart et al., 2013).

It is interesting to note that investigations show clearly the importance of physical activity and a healthy lifestyle; however, few people can be convinced to participate in a long study, 7% of people who were asked showed any interest (Atlantis et al., 2006). This might indicate that genes that enable the body to save and accumulate energy, also known as "laziness genes," are active (Chakravarthy and Booth, 2004). Thus, awareness of and pleasure in exercising are important to make physical exercise a habit. It is known that physical training increases the level of endocannabinoids and induce euphoria and satisfaction (see Chapter 3, "R" Extra).

9.2 DOES HIGH VO$_2$MAX DECREASES THE INCIDENCE OF CANCER?

Several epidemiologic studies imply that physical fitness decreases the incidence of some type of cancer (Umar et al., 2012). Blair et al. (1995) have shown that physical fitness prevented the development of cardiovascular diseases and cancer, and physical fitness was strongly determined by the level of physical wellness. Sawada and colleagues have shown that higher VO_2max decreased the incidence of cancer significantly (Sawada et al., 2003). It is not generally true for any type of cancer—for instance, breast cancer can be prevented by physical training; exercising is beneficial to endocrine functions and alters receptors involved in cancerous signaling (Bernstein, 2009; Charkoudian and Joyner, 2004). Although physical training and high VO_2max seem to prevent breast cancer, the mechanisms are still elusive.

A meta-analysis including more than 88,000 individuals showed that prostate cancer, which is a common cancer type in males, was uncommon among those who trained frequently (Liu et al., 2011). The mechanism is not understood yet, but physical activity raising antioxidant levels to ameliorate oxidative stress might be one factor; additionally, physical activity alters metabolism (Antonelli et al., 2009; Rebillard et al., 2013; Sea et al., 2009). Our group and others have observed that physical activity slows the enlargement of tumors, also in animals with prostate cancer graft (Esser et al., 2009; Radak, 2004; Radak et al., 2001b, 2002). Colon cancer is also altered by physical activity. A recent meta-analysis including 21 individuals showed 27% lower incidence in individuals who trained frequently (Boyle et al., 2012). The preventive effect applies to both types located in the proximal and distal ends of the colon. An explanation might be that physical activity influences peristaltic movements, thus elimination of toxic components is faster. A surprising result has been shown in animals exposed to UV radiation in a recent study. Skin cancer was less frequent in those animals that consumed caffeine and were allowed to run on a wheel. The underlying cause might be the decreased fat content under the skin and changes in programmed cell death (Conney et al., 2013).

The most important factor in fighting against cancer is the reinforcement of the immune system, and frequent physical training has a beneficial effect on this. The immune system is a complex system including innate and adaptive defense processes. The relationship between physical fitness and immune responses can described by a U-shaped curve, with decreased efficiency of the immune system in undertrained and overtrained individuals, which presents an increased risk (Fig. 9.3).

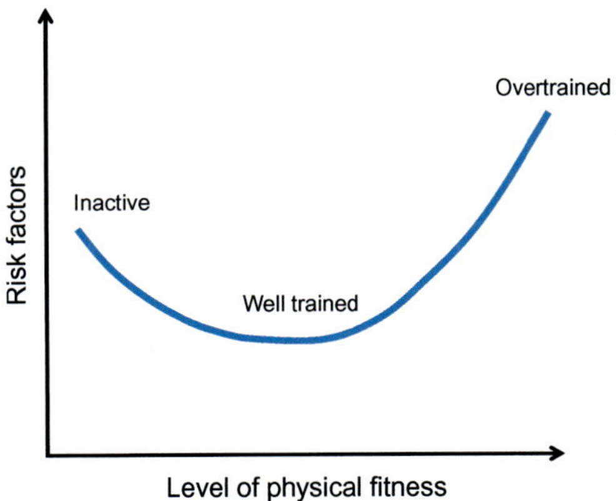

Fig. 9.3 Physical fitness and lifestyle-related diseases. The hypothetical relationship between physical fitness and lifestyle-related diseases assumes an increased risk in undertrained and overtrained individuals.

Inflammation is a protective mechanism against harmful stimuli and injuries. It is protective because the increased blood flow following cellular injury (mechanical, chemical, radiation) promotes elimination of toxic components and damaged cells, thus it does not spread out and affect a larger area. Thus, inflammation is an important protective, reparative mechanism, which is regulated by proinflammatory and antiinflammatory cytokines. Chronic inflammation is an avoidable condition since it promotes not regeneration but damage to cells, which is mostly oxidative damage. Inflammation is a common phenomenon during physical training, and its localization and amelioration are the most important elements in treating sport micro and macro injuries. Cooling is a useful method following high loading training to minimize inflammation. It is interesting to note that physical training has been in the spotlight of preventing immune-related diseases such as rheumatic arthritis and Alzheimer's disease. It has been shown that frequent physical training suppresses or reduces inflammation through NF-κB, which is a transcription factor that regulates immune processes (Radak et al., 2004; Toldy et al., 2009). It should be noted that high intensity training for a couple of hours impairs immune functions. This phenomenon is called the "open window," and, for instance, upper respiratory tract infection may occur during this short period.

In summary, observations imply that frequent physical training decreases the incidence of some types of cancer. This might be a complex mechanism including hormonal alterations, intestinal peristaltic movement, and programmed cell death. It is also an important observation that physical training improves the efficiency of the immune system.

9.3 VO₂MAX AND NEURODEGENERATION

Low VO$_2$max increases the risk of the development of cardiovascular diseases, metabolic disorders, cancer, and neurodegenerative diseases. Age affects the incidence of stroke, Alzheimer's and Parkinson's diseases, while physical activity seems to prevent these disturbances (Radak et al., 2007, 2010). Frequent training increases the size of the hippocampus (Erickson et al., 2011; Fotuhi et al., 2012), which correlates with an improved memory and learning capacity following training. One well-described mechanism, by which physical training exerts its influence on the brain, is the brain-derived neurotrophic factor (BDNF) pathway. BDNF plays a role in synaptic plasticity, metabolic processes, neurogenesis, and stress resistance of neuronal cells (Rothman et al., 2012), and physical training also has an antidepressant effect. Exercising stimulates nerve growth factors (NGF), which play a role in axonal branching and elongation (Madduri et al., 2009), and which have been observed following training (Doyle and Roberts, 2006). Increased oxidative damage adversely affects brain functions (Butterfield et al., 1997), and physical training lessens the extent of oxidative damage and ameliorates brain aging processes and improves cognitive functions in animals (Radak et al., 2001a). Furthermore, exercising activates oxidative damage-repairing enzymes (Koltai et al., 2011) and stimulates the activity of "housekeeping" enzymes such as proteasomes, which may play a role in reducing oxidative damage (Radak et al., 2001a, 2011). Housekeeping enzymes degrade damaged useless proteins and repair DNA damage. Physical training also stimulates the activity of neprilysin and insulin-degrading enzymes, which also degrade beta-amyloid, and this protein accumulates in the spaces between neurons in Alzheimer's disease (Maesako et al., 2012; Radak et al., 2010). In addition, exercising improves blood supply to the brain and vascularization (Tang et al., 2010), which may play a role in the regulation of the metabolic activity of the brain (Lou et al., 2008). These alterations can be followed in human samples also in addition to laboratory animals (Bullitt et al., 2009). Modern imaging methods allow the investigation of the effects of physical training on aging processes and structural deterioration (Colcombe et al., 2006). It has been demonstrated that exercising decreases the extent of brain softening, reduces the loss of brain tissue in the elderly, and lessens the thinning of cortex compared to elderly individuals who are physically inactive. VO_2max (cardiovascular fitness) strongly correlates with learning in human individuals (Holzschneider et al., 2012) and also in animals (Sarga et al., 2013) (Fig. 9.4).

9.4 METABOLIC SYNDROME AND TYPE 2 DIABETES

Obesity, metabolic syndrome, and type 2 diabetes are common diseases of the 21st century affecting millions of individuals in the world, and they are likely to increase. Quantity and quality of food and the amount that is used by the body influence the development of diseases. Thus, a well-designed diet and physical training have a huge impact on health and reduce the incidence of metabolic disorders.

Chapter 8 contained detailed discussion about a well-designed diet, and this chapter address the effects of physical training. Metabolic syndrome is a clustering of metabolic conditions such as obesity, significant increase of abdominal circumference, high triglyceride and LDL levels, high blood pressure, and higher fasting blood sugar level and insulin resistance (Fig. 9.5).

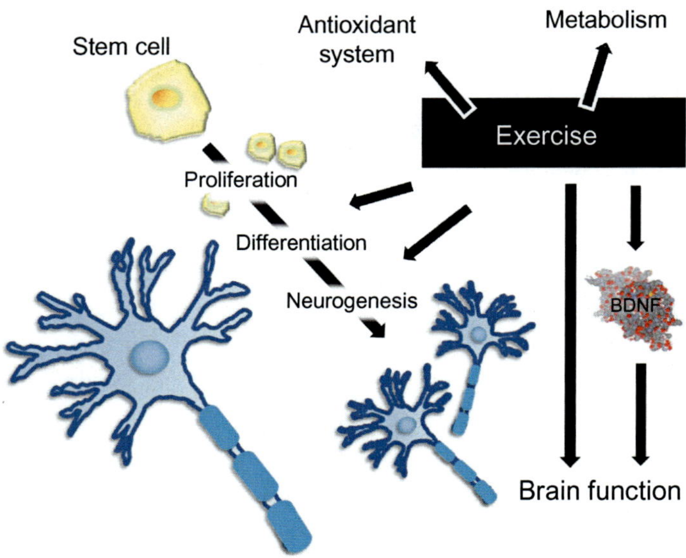

Fig. 9.4 Physical training and brain functions. Physical training improves brain functions via several cellular mechanisms.

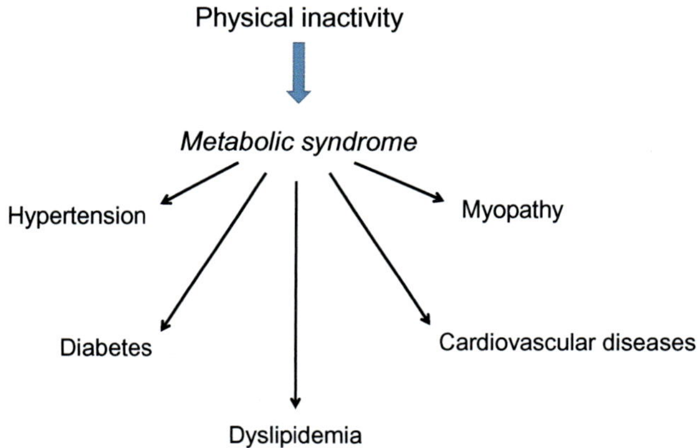

Fig. 9.5 Metabolic syndrome. The consequences of metabolic syndrome are presented here, and its main risk factor is physical inactivity. Frequent physical training is the most important tool to prevent and treat metabolic syndrome.

Abdominal circumference can be measured easily, and its high value is associated with higher risk of cardiovascular diseases. Metabolic syndrome may result in myositis and muscle atrophy, which is also called myopathy. These processes are partly free radical induced mechanisms, and inflammation in the muscle promotes degradation to a higher extent; thus catabolic processes overbalance anabolic processes.

Metabolic syndrome correlates strongly with the development of several diseases including cardiovascular diseases and type 2 diabetes. Frequent physical training decreases the chances of developing metabolic syndrome, and the extent of loading and duration of exercising are critical factors: at least 60 min a day with moderate intensity (Strasser, 2013). Moderate intensity is about 3–6 MET (metabolic equivalent of task) according to the American College of Sport Medicine (ACSM). MET expresses the energy coast of physical activities, and is defined as a ratio of metabolic rate during a specific physical activity to a reference metabolic rate. Oxygen consumption of 3.5 mL/kg/min is equivalent to 1 MET = 4.184 kJ/kg/h energy expenditure. Resting or sitting consume 1 MET, walking at a pace of 4 km/h is 2.9 MET, jogging 6 MET, and intensive jump rope workout 10 MET. Thus, MET is calculated from values of the average population, and due to significant individual differences, its scientific accuracy is not always sufficient. However, it is a very useful measure in epidemiological studies. One study has shown that frequency, quality, and quantity of physical activity are inversely correlated with obesity; active individuals had normal body weight, while inactive individuals were overweight (Healy et al., 2008). It has also been demonstrated that obesity is associated with an increased triglyceride level (Healy et al., 2008). There was statistical correlation between physical inactivity measured by duration of resting and sitting (e.g., watching TV, driving a car, etc.), and obesity and type 2 diabetes (Hu et al., 2003; Katzmarzyk et al., 2009). The preventive effects of exercising are well understood since the energy demands of basic metabolic processes are constant, and the total energy consumption of the body can be enhanced by physical activity, thereby preventing the accumulation of fat. The energy demand of physical activity can be significant—for instance, Tour de France athletes consume 8000 kcal a day.

Physical training also prevents the development of high blood pressure (Tschentscher et al., 2013). The underlying mechanism behind it includes the increased blood flow and consequential increased shear stress on the endothelial cells, resulting in activation of nitrogen oxide synthase (eNOS), NO production, and consequent vascular relaxation, thus blood pressure decreases. Frequent physical training leads to chronically high levels of eNOS, and vascular elasticity can be preserved.

An Australian investigation on the daily activities of 8000 individuals suffering from type 2 diabetes revealed that physical inactivity, resting, and sitting show strong correlations with obesity and diabetes (Dunstan et al., 2007). Insulin-dependent glucose uptake mainly occurs through skeletal muscle.

Food intake increases blood sugar level, which is a healthy physiological phenomenon, and this stimulates pancreatic beta cells to produce insulin. Insulin is then released into the bloodstream and travels to many organs and binds to cell surface insulin receptors (IR). Upon binding to skeletal muscle, intracellular glucose transporters GLUT4 are translocated to the membrane and allow glucose uptake by the cell. This is the mechanism of action of insulin. Skeletal muscle and liver are the two main insulin-dependent glucose consumers of the body; glucose uptake by the brain is insulin-independent. Glucose can be converted into glycogen or fat in skeletal muscle. Frequent contractions of skeletal muscle consume glucose continuously, preventing glucose conversion to fat, whereas inactivity allows the accumulation of fat. Fat is not able to uptake glucose, and the decreased muscle mass ratio impairs glucose consumption. It activates compensation mechanisms that produce higher amounts of insulin; however, it is associated with impaired glucose uptake. The underlying factors include reduced amount of IRs and their efficiency, reduced amount of GLUT4, reduced muscle mass, increased fat tissue, and physical inactivity. This phenomenon is called insulin resistance, and it is the main cause of type 2 diabetes. Frequent physical training decreases the amount of fat, and increases muscle mass and GLUT4. These are the main factors why physical training is the most effective tool in preventing type 2 diabetes. It is important to note that individuals suffering from type 2 diabetes benefit from exercising, since this also reduces glycation due to high blood glucose levels. Glycation affects proteins since glucose molecules are attached to them, thereby influencing protein structure, which adversely affects their function. Insulin resistance is ameliorated in individuals with type 2 diabetes following physical training, thus less insulin might be sufficient. Type 2 diabetes leads to diabetic complications, which affects the function of the kidney and eye, and results in cardiovascular diseases (Fig. 9.5).

9.5 "R" EXTRA

A huge amount of microorganisms live in the human body, their number exceeding maybe 10 times the number of human cells (Gill et al., 2006). They are located mostly in the intestinal tract, and recent discoveries have shown the importance of gut flora in health, since these microorganisms are able to induce inflammatory processes, alter immune functions and enlargement of a tumor, and they may play a role in obesity and cardiovascular disorders (Grenham et al., 2011). Physical training also influences gut flora and is thought to play a role in prevention, which has been observed in cases of colon cancer (Matsumoto et al., 2008). Frequent running exercises increased the level of intestinal *n*–butyrate in mice, a by-product of metabolic processes of symbiotic microorganisms. *n*-Butyrate is a short chain fatty acid, which suppress inflammatory processes and oxidative stress, and regulates endothelial permeability in the intestines (Canani et al., 2011). Additionally, lower levels of free radical production results in lower levels of DNA damage and mutations, which may associated with lower chances of developing cancer. The microbiome is also linked to obesity, different type of cancer, and neurodegenerative diseases, like Alzheimer's disease. It appears that regular exercise has the capability to alter levels of certain microorganisms in the microbiome, and by this it further regulates the body.

9.6 SUMMARY

Physical inactivity is a risk factor of almost every disease, thus frequent training has significant preventive effects. Health prevention means that healthy individuals increase their resistance to certain disorders, and maintain their balance during such external conditions that induce pathological processes in untrained individuals. This is the point of prevention, and physical training is the cheapest and most effective way to maintain health. Lifestyle-related diseases can be prevented efficiently by frequent physical exercise. Cardiovascular fitness (VO_2max value) is a useful biomarker of health; its value shows an inverse correlation with the development of cardiovascular and heart diseases, certain types of cancer, and metabolic disorders.

TEST QUESTIONS

1. What is the relationship between cardiovascular fitness and prevention?
2. How can you describe the exercise-associated prevention of diseases?
3. How does frequent physical training affect the development of cancer?
4. What are the main effects of physical training on the nervous system?
5. How does physical training affect type 2 diabetes?
6. How does physical training influence blood pressure?
7. What are the underlying factors behind insulin resistance?
8. What are the characteristics of metabolic syndrome?
9. What is MET?

BIBLIOGRAPHY

Andersen, L.B., 1995. Physical activity and physical fitness as protection against premature disease or death. Scand. J. Med. Sci. Sports 5, 318–328.

Antonelli, J.A., Jones, L.W., Banez, L.L., Thomas, J.A., Anderson, K., Taylor, L.A., Gerber, L., Anderson, T., Hoyo, C., Grant, D., Freedland, S.J., 2009. Exercise and prostate cancer risk in a cohort of veterans undergoing prostate needle biopsy. J. Urol. 182, 2226–2231.

Atlantis, E., Chow, C.M., Kirby, A., Fiatarone Singh, M.A., 2006. Worksite intervention effects on physical health: a randomized controlled trial. Health Promot. Int. 21, 191–200.

Bernstein, L., 2009. Exercise and breast cancer prevention. Curr. Oncol. Rep. 11, 490–496.

Blair, S.N., Kohl 3rd, H.W., Barlow, C.E., Paffenbarger Jr., R.S., Gibbons, L.W., Macera, C.A., 1995. Changes in physical fitness and all-cause mortality. A prospective study of healthy and unhealthy men. JAMA 273, 1093–1098.

Boyle, T., Keegel, T., Bull, F., Heyworth, J., Fritschi, L., 2012. Physical activity and risks of proximal and distal colon cancers: a systematic review and meta-analysis. J. Natl. Cancer Inst. 104, 1548–1561.

Bramble, D.M., Lieberman, D.E., 2004. Endurance running and the evolution of Homo. Nature 432, 345–352.

Brien, S.E., Katzmarzyk, P.T., Craig, C.L., Gauvin, L., 2007. Physical activity, cardiorespiratory fitness and body mass index as predictors of substantial weight gain and obesity: the Canadian physical activity longitudinal study. Can. J. Public Health 98, 121–124.

Bullitt, E., Rahman, F.N., Smith, J.K., Kim, E., Zeng, D., Katz, L.M., Marks, B.L., 2009. The effect of exercise on the cerebral vasculature of healthy aged subjects as visualized by MR angiography. AJNR Am. J. Neuroradiol. 30, 1857–1863.

Butterfield, D.A., Howard, B.J., Yatin, S., Allen, K.L., Carney, J.M., 1997. Free radical oxidation of brain proteins in accelerated senescence and its modulation by N-tert-butyl-alpha-phenylnitrone. Proc. Natl. Acad. Sci. U. S. A. 94, 674–678.

Canani, R.B., Costanzo, M.D., Leone, L., Pedata, M., Meli, R., Calignano, A., 2011. Potential beneficial effects of butyrate in intestinal and extraintestinal diseases. World J. Gastroenterol. 17, 1519–1528.

Chakravarthy, M.V., Booth, F.W., 2004. Eating, exercise, and "thrifty" genotypes: connecting the dots toward an evolutionary understanding of modern chronic diseases. J. Appl. Physiol. 96, 3–10.

Charkoudian, N., Joyner, M.J., 2004. Physiologic considerations for exercise performance in women. Clin. Chest Med. 25, 247–255.

Colcombe, S.J., Erickson, K.I., Scalf, P.E., Kim, J.S., Prakash, R., McAuley, E., Elavsky, S., Marquez, D.X., Hu, L., Kramer, A.F., 2006. Aerobic exercise training increases brain volume in aging humans. J. Gerontol. A Biol. Sci. Med. Sci. 61, 1166–1170.

Conney, A.H., Lou, Y.R., Nghiem, P., Bernard, J.J., Wagner, G.C., Lu, Y.P., 2013. Inhibition of UVB-induced nonmelanoma skin cancer: a path from tea to caffeine to exercise to decreased tissue fat. Top. Curr. Chem. 329, 61–72.

Doyle, L.M., Roberts, B.L., 2006. Exercise enhances axonal growth and functional recovery in the regenerating spinal cord. Neuroscience 141, 321–327.

Dunstan, D.W., Salmon, J., Healy, G.N., Shaw, J.E., Jolley, D., Zimmet, P.Z., Owen, N., 2007. Association of television viewing with fasting and 2-h postchallenge plasma glucose levels in adults without diagnosed diabetes. Diabetes Care 30, 516–522.

Dvorak, R.V., Tchernof, A., Starling, R.D., Ades, P.A., DiPietro, L., Poehlman, E.T., 2000. Respiratory fitness, free living physical activity, and cardiovascular disease risk in older individuals: a doubly labeled water study. J. Clin. Endocrinol. Metab. 85, 957–963.

Erickson, K.I., Voss, M.W., Prakash, R.S., Basak, C., Szabo, A., Chaddock, L., Kim, J.S., Heo, S., Alves, H., White, S.M., Wojcicki, T.R., Mailey, E., Vieira, V.J., Martin, S.A., Pence, B.D., Woods, J.A., McAuley, E., Kramer, A.F., 2011. Exercise training increases size of hippocampus and improves memory. Proc. Natl. Acad. Sci. U. S. A. 108, 3017–3022.

Esser, K.A., Harpole, C.E., Prins, G.S., Diamond, A.M., 2009. Physical activity reduces prostate carcinogenesis in a transgenic model. Prostate 69, 1372–1377.

Fotuhi, M., Do, D., Jack, C., 2012. Modifiable factors that alter the size of the hippocampus with ageing. Nat. Rev. Neurol. 8, 189–202.

Gill, S.R., Pop, M., Deboy, R.T., Eckburg, P.B., Turnbaugh, P.J., Samuel, B.S., Gordon, J.I., Relman, D.A., Fraser-Liggett, C.M., Nelson, K.E., 2006. Metagenomic analysis of the human distal gut microbiome. Science 312, 1355–1359.

Grenham, S., Clarke, G., Cryan, J.F., Dinan, T.G., 2011. Brain-gut-microbe communication in health and disease. Front. Physiol. 2, 94.

Hart, N., Sarga, L., Csende, Z., Koltai, E., Koch, L.G., Britton, S.L., Davies, K.J., Kouretas, D., Wessner, B., Radak, Z., 2013. Resveratrol enhances exercise training responses in rats selectively bred for high running performance. Food Chem. Toxicol. 61, 53–59.

Healy, G.N., Wijndaele, K., Dunstan, D.W., Shaw, J.E., Salmon, J., Zimmet, P.Z., Owen, N., 2008. Objectively measured sedentary time, physical activity, and metabolic risk: the Australian diabetes, obesity and lifestyle study (AusDiab). Diabetes Care 31, 369–371.

Holzschneider, K., Wolbers, T., Roder, B., Hotting, K., 2012. Cardiovascular fitness modulates brain activation associated with spatial learning. Neuroimage 59, 3003–3014.

Hu, F.B., Li, T.Y., Colditz, G.A., Willett, W.C., Manson, J.E., 2003. Television watching and other sedentary behaviors in relation to risk of obesity and type 2 diabetes mellitus in women. JAMA 289, 1785–1791.

Hussain, S.O., Barbato, J.C., Koch, L.G., Metting, P.J., Britton, S.L., 2001. Cardiac function in rats selectively bred for low- and high-capacity running. Am. J. Physiol. Regul. Integr. Comp. Physiol. 281, R1787–1791.

Katzmarzyk, P.T., Church, T.S., Craig, C.L., Bouchard, C., 2009. Sitting time and mortality from all causes, cardiovascular disease, and cancer. Med. Sci. Sports Exerc. 41, 998–1005.

Koch, L.G., Kemi, O.J., Qi, N., Leng, S.X., Bijma, P., Gilligan, L.J., Wilkinson, J.E., Wisloff, H., Hoydal, M.A., Rolim, N., Abadir, P.M., van Grevenhof, E.M., Smith, G.L., Burant, C.F., Ellingsen, O., Britton, S.L., Wisloff, U., 2011. Intrinsic aerobic capacity sets a divide for aging and longevity. Circ. Res. 109, 1162–1172.

Koch, L.G., Britton, S.L., Wisloff, U., 2012. A rat model system to study complex disease risks, fitness, aging, and longevity. Trends Cardiovasc. Med. 22, 29–34.

Koltai, E., Zhao, Z., Lacza, Z., Cselenyak, A., Vacz, G., Nyakas, C., Boldogh, I., Ichinoseki-Sekine, N., Radak, Z., 2011. Combined exercise and insulin-like growth factor-1 supplementation induces neurogenesis in old rats, but do not attenuate age-associated DNA damage. Rejuvenation Res. 14, 585–596.

Laukkanen, J.A., Laaksonen, D., Lakka, T.A., Savonen, K., Rauramaa, R., Makikallio, T., Kurl, S., 2009. Determinants of cardiorespiratory fitness in men aged 42 to 60 years with and without cardiovascular disease. Am. J. Cardiol. 103, 1598–1604.

Liu, Y., Hu, F., Li, D., Wang, F., Zhu, L., Chen, W., Ge, J., An, R., Zhao, Y., 2011. Does physical activity reduce the risk of prostate cancer? A systematic review and meta-analysis. Eur. Urol. 60, 1029–1044.

Lou, S.J., Liu, J.Y., Chang, H., Chen, P.J., 2008. Hippocampal neurogenesis and gene expression depend on exercise intensity in juvenile rats. Brain Res. 1210, 48–55.

Madduri, S., Papaloizos, M., Gander, B., 2009. Synergistic effect of GDNF and NGF on axonal branching and elongation in vitro. Neurosci. Res. 65, 88–97.

Maesako, M., Uemura, K., Kubota, M., Kuzuya, A., Sasaki, K., Hayashida, N., Asada-Utsugi, M., Watanabe, K., Uemura, M., Kihara, T., Takahashi, R., Shimohama, S., Kinoshita, A., 2012. Exercise is more effective than diet control in preventing high fat diet-induced beta-amyloid deposition and memory deficit in amyloid precursor protein transgenic mice. J. Biol. Chem. 287, 23024–23033.

Matsumoto, M., Inoue, R., Tsukahara, T., Ushida, K., Chiji, H., Matsubara, N., Hara, H., 2008. Voluntary running exercise alters microbiota composition and increases n-butyrate concentration in the rat cecum. Biosci. Biotechnol. Biochem. 72, 572–576.

Radak, Z. (Ed.), 2004. Exercise and Diseases: Prevention through Training. Meyer and Meyer Sport, Aachen.

Radak, Z., Kaneko, T., Tahara, S., Nakamoto, H., Pucsok, J., Sasvari, M., Nyakas, C., Goto, S., 2001a. Regular exercise improves cognitive function and decreases oxidative damage in rat brain. Neurochem. Int. 38, 17–23.

Radak, Z., Taylor, A.W., Sasvari, M., Ohno, H., Horkay, B., Furesz, J., Gaal, D., Kanel, T., 2001b. Telomerase activity is not altered by regular strenuous exercise in skeletal muscle or by sarcoma in liver of rats. Redox Rep. 6, 99–103.

Radak, Z., Gaal, D., Taylor, A.W., Kaneko, T., Tahara, S., Nakamoto, H., Goto, S., 2002. Attenuation of the development of murine solid leukemia tumor by physical exercise. Antioxid. Redox Signal. 4, 213–219.

Radak, Z., Chung, H.Y., Naito, H., Takahashi, R., Jung, K.J., Kim, H.J., Goto, S., 2004. Age-associated increase in oxidative stress and nuclear factor kappaB activation are attenuated in rat liver by regular exercise. FASEB J. 18, 749–750.

Radak, Z., Kumagai, S., Taylor, A.W., Naito, H., Goto, S., 2007. Effects of exercise on brain function: role of free radicals. Appl. Physiol. Nutr. Metab. 32, 942–946.

Radak, Z., Hart, N., Sarga, L., Koltai, E., Atalay, M., Ohno, H., Boldogh, I., 2010. Exercise plays a preventive role against Alzheimer's disease. J. Alzheimers Dis. 20, 777–783.

Radak, Z., Zhao, Z., Goto, S., Koltai, E., 2011. Age-associated neurodegeneration and oxidative damage to lipids, proteins and DNA. Mol. Aspects Med. 32, 305–315.

Rebillard, A., Lefeuvre-Orfila, L., Gueritat, J., Cillard, J., 2013. Prostate cancer and physical activity: adaptive response to oxidative stress. Free Radic. Biol. Med. 60, 115–124.

Rothman, S.M., Griffioen, K.J., Wan, R., Mattson, M.P., 2012. Brain-derived neurotrophic factor as a regulator of systemic and brain energy metabolism and cardiovascular health. Ann. N.Y. Acad. Sci. 1264, 49–63.

Sarga, L., Hart, N., Koch, L.G., Britton, S.L., Hajas, G., Boldogh, I., Ba, X., Radak, Z., 2013. Aerobic endurance capacity affects spatial memory and SIRT1 is a potent modulator of 8-oxoguanine repair. Neuroscience 252, 326–336.

Sawada, S.S., Muto, T., Tanaka, H., Lee, I.M., Paffenbarger Jr., R.S., Shindo, M., Blair, S.N., 2003. Cardiorespiratory fitness and cancer mortality in Japanese men: a prospective study. Med. Sci. Sports Exerc. 35, 1546–1550.

Sea, J., Poon, K.S., McVary, K.T., 2009. Review of exercise and the risk of benign prostatic hyperplasia. Physician Sports Med. 37, 75–83.

Strasser, B., 2013. Physical activity in obesity and metabolic syndrome. Ann. N.Y. Acad. Sci. 1281, 141–159.

Tang, K., Xia, F.C., Wagner, P.D., Breen, E.C., 2010. Exercise-induced VEGF transcriptional activation in brain, lung and skeletal muscle. Respir. Physiol. Neurobiol. 170, 16–22.

Toldy, A., Atalay, M., Stadler, K., Sasvari, M., Jakus, J., Jung, K.J., Chung, H.Y., Nyakas, C., Radak, Z., 2009. The beneficial effects of nettle supplementation and exercise on brain lesion and memory in rat. J. Nutr. Biochem. 20, 974–981.

Totsikas, C., Rohm, J., Kantartzis, K., Thamer, C., Rittig, K., Machann, J., Schick, F., Hansel, J., Niess, A., Fritsche, A., Haring, H.U., Stefan, N., 2011. Cardiorespiratory fitness determines the reduction in blood pressure and insulin resistance during lifestyle intervention. J. Hypertens. 29, 1220–1227.

Tschentscher, M., Niederseer, D., Niebauer, J., 2013. Health benefits of Nordic walking: a systematic review. Am. J. Prev. Med. 44, 76–84.

Umar, A., Dunn, B.K., Greenwald, P., 2012. Future directions in cancer prevention. Nat. Rev. Cancer 12, 835–848.

Williams, P.T., 2001. Physical fitness and activity as separate heart disease risk factors: a meta-analysis. Med. Sci. Sports Exerc. 33, 754–761.

Wisloff, U., Najjar, S.M., Ellingsen, O., Haram, P.M., Swoap, S., Al-Share, Q., Fernstrom, M., Rezaei, K., Lee, S.J., Koch, L.G., Britton, S.L., 2005. Cardiovascular risk factors emerge after artificial selection for low aerobic capacity. Science 307, 418–420.

Zimmer, C., 2004. Human evolution. Faster than a hyena? Running may make humans special. Science 306, 1283.

CHAPTER 10

Physical Training and Aging

10.1 INTRODUCTION

Aging is a decline in autonomic processes of bodily functions due to primary alterations in genetic regulation. Aging is defined in many ways; the most popular is that aging is a decline in bodily ability to maintain homeostasis. Homeostasis keeps physiological functions in an optimal range, which enables the body to respond effectively to external challenges. Thus, adaptability of the body is at its most advanced when homeostasis working well.

It is well-known that adaptability declines significantly during aging (Radak et al., 2008a). Excellent evolutionary adaptation of *Homo sapiens* was the main success factor during thousands of years of human evolution. One way to improve adaptation ability is frequent physical training. Physical activity means sequential loading (training–resting), which promotes adaptation processes in every organs.

Aging affect the whole body; however, aging of human organs varies, thus finding only one biomarker seems like an utopistic conception (e.g., aging of eye differs from bone thinning or aging of the kidney). When considering age, average life expectancy and maximal lifespan should be differentiated. The human maximal lifespan is 122 years—a French lady (Jeanne Calment) lived longest, according to records. Average life expectancy depends on many factors including economic growth and gross domestic product (GDP); it is higher in developed countries. Cuban nationals live longer than expected based on their economic circumstances, whereas Hungarians do not live as long as could be expected (Fig. 10.1).

Life expectancy is influenced by sex; females live longer than males. It is also determined by lifestyle, diet, and physical fitness, since these all influence the development of certain diseases and thereby life expectancy (Fig. 10.2).

This chapter summarizes the factors by which physical training exerts its influence on aging. Maximal lifespan is not altered by physical training according to recent knowledge; however, exercising significantly alters factors that play a role in aging such as amount of free radicals and the accumulation of oxidative damage caused by them, and also alteration of IGF-1 signaling pathway, sirtuins, and metabolic changes.

The Physiology of Physical Training
https://doi.org/10.1016/B978-0-12-815137-2.00010-3
 157

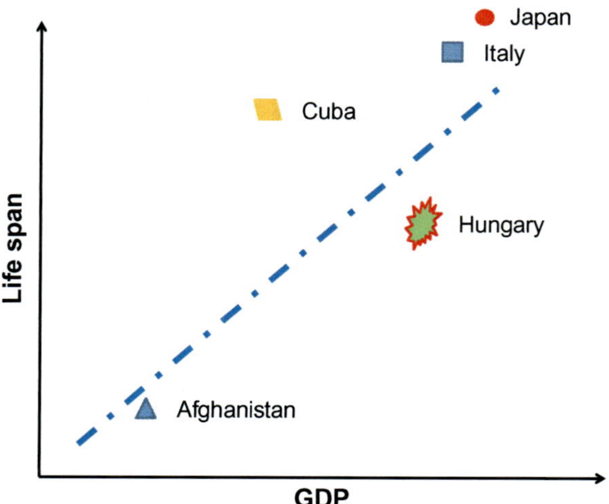

Fig. 10.1 GDP and average life expectancy. Gross domestic product (GDP) *per capita*, which is a measure of economic activity within a country, shows a strong correlation with average life expectancy.

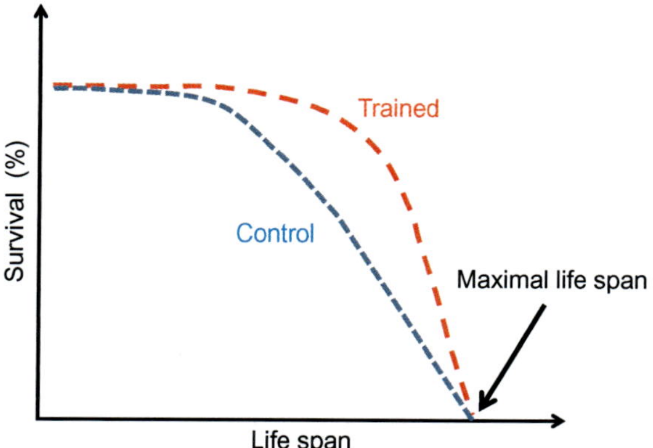

Fig. 10.2 Effect of physical fitness on life expectancy. Frequent physical activity increases average life expectancy compared to an inactive lifestyle, as observed in both human individuals and laboratory animals. Graph based on data extracted from scientific literature.

10.2 HYPOTHETIC MODELS OF AGING

There are two main aging concepts represented by scientists: one group thinks that aging is determined by only one factor; the other group thinks that aging is a complex process determined by many factors. More than 300 aging concepts have been used to attempt to explain aging processes. This chapter addresses the most important ones.

10.2.1 Error Catastrophe Theory of Aging

Leo Szilárd, an outstanding Hungarian-born physicist, proposed that aging is a result of accumulation of mutation of somatic cells of the body, and therefore information encoded in the DNA cannot be applied, which leads to a catastrophe in the cell (Szilard, 1959). However, there has been no definite proof that this is the case. This theory established the foundation of theories that focus on the alteration of DNA sequence.

10.2.2 Telomere Theory

Hayflick's concept of aging (Linskens et al., 1995), based on the length of telomeres, implies that aging is a programmed process and set by a "genetic clock." The clock is the telomere cap at the end of the chromosomes, which shortens with each cell division (Fig. 10.3).

When a telomere reaches a certain length, diploid cells cease to divide; this is called cellular senescence. The length of the telomere is regulated by telomerase enzyme, which is active in fetal tissue and does not allow shortening of the telomere. At the end of embryonic development, its activity is lost, resulting in limited cell divisions (about 50). If telomerase regain its activity, cells become immortal, which is the case in cancerous cells. This theory is one of the the most accepted aging concepts.

10.2.3 Sirtuin Theory

This theory emphasizes the role of genetic factors in aging. Sirtuin proteins play a role in aging; caloric restriction (CR) has been demonstrated to increase SIR2 gene expression in yeast, and is associated with lifespan. Increasing the amount of SIR2 resulted in a longer lifespan, and if this gene

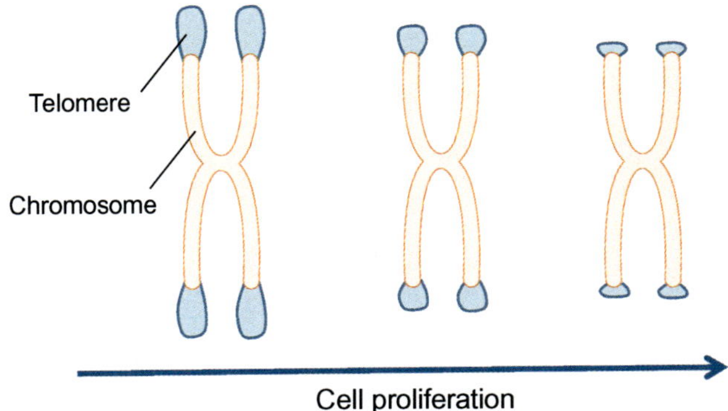

Fig. 10.3 Telomere length shortening. Telomere shortening with each cell division limits the number a cell can divide.

were deleted, it resulted in a shorter lifespan. Mammals possess seven iso-forms of SIRT proteins, which are NAD-dependent proteins, thereby they are sensitive markers of cellular metabolism (Radak et al., 2013). The main function of SIRT proteins is deacetylation of lysine amino acids in histones and other proteins including transcription factors (Fig. 10.4).

Deacetylation of lysine in histones makes the chromosomes' structure more compact, and genes are less available for transcription; this reduces the extent of adaptation responses to different stimuli. However, a more compact chromatic structure protects DNA against damage, which could result in apoptotic cell death. Thus, sirtuins attenuate apoptosis via deacetylation of histones, so they are called antiapoptotic proteins. Adaptation capacity declines with age. Several investigations have suggested that activation of SIRT proteins by CR, physical training, or resveratrol may result in longevity and better quality of life. Sirtuins are able to deacetylate lysine amino acids in any proteins, SIRT1 deacetylate p53t, FOXO groups, NF-κB, PGC-1 alpha and other transcription factors that regulates expression and activity of antioxidants, inflammation regulators, and enzymes involved in the repair of oxidative damage. The effects of SIRT proteins are complex; they regulate metabolic processes at several levels.

10.2.4 Wear and Tear Theory

The length and quality of teeth of animals deteriorate with age, as do other organs. This observation led to the theory of wear and tear, where certain parts of organs are worn away with age, e.g., the wings of flies, joints in the human body. This theory covers those organs that have limited options to be renewed, such as teeth. Regeneration of articular cartilage in joints is limited.

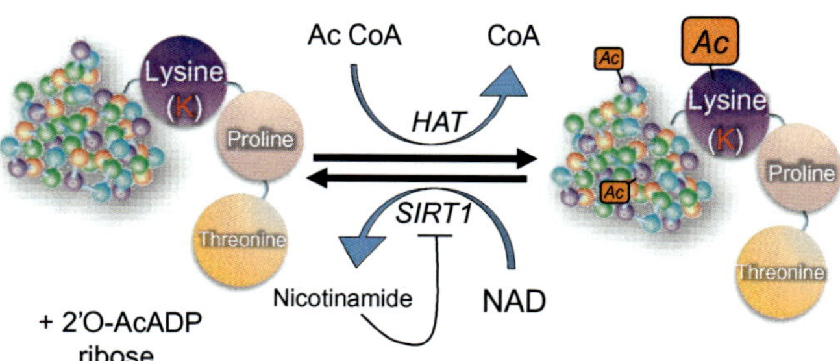

Fig. 10.4 Acetylation and deacetylation. Sirtuin enzyme removes the acetyl group from lysine amino acids (K) in proteins by consuming NAD. This protein family plays a role in metabolic processes, histone regulation, and DNA repair, thereby regulating aging processes.

10.2.5 Free Radicals Theory

From the mid-1950s, molecules with unpaired electrons were suspected to react easily and attack any cellular organelles. Fatty acid peroxidation and oxidative damage of proteins and DNA increase significantly in the last third of life, and these occurrences are associated with a decline in physiological functions. The free radicals theory of aging were introduced by Demon Harman in 1956 (Harman, 1956), and it emphasizes that aging is a free radicals-dependent process. This theory is the most acceptable concept of aging, and several investigations support this theory. More free radicals are produced with age since the activity of enzymes that generates free radicals (NADPH-oxidase, xanthine-oxidase) increases, which results in elevated oxidative damage. Antioxidant defense mechanisms are impaired with age, and the activity of repair enzymes declines as well. In addition, the activity of housekeeping enzymes is also impaired; these enzymes are responsible for elimination of damaged proteins and repairing cellular oxidative damage. Accumulated oxidative damage impairs cellular functions and induces chain reactions that further stimulate free radical generation (Radak et al., 2008b) (Fig. 10.5).

10.2.6 Membrane Theory

Membrane theory, established by Istvan Zs. Nagy, emphasizes the role of altered membrane fluidity in aging, which results in impaired water-storing ability. Indeed, cellular liquid content does decline with age. This can be clearly recognized in the skin. The question is why membrane fluidity changes. The role of free radicals-related lipid peroxidation seems to be the answer.

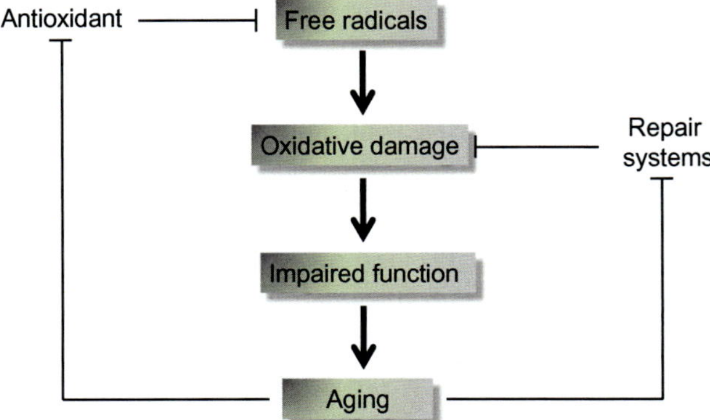

Fig. 10.5 Free radical-induced damage in macromolecules. The speculative relationship between free radicals-induced damage in macromolecules and aging.

10.2.7 Apoptosis Theory (Cell Death Upon Genetic Crisis)

Dividing cells renew continuously; however, errors may occur during protein synthesis and improper replication of DNA during mitochondrial biogenesis. Errors originated from synthesis, normal cellular functioning, or from injury mostly caused by free radicals cannot be passed to the next generation of cells, because these errors would put the organism at risk of mutation. In order to avoid such a risk of transmitting errors and mutations, these erroneous cells commit suicide via a mechanism called programmed cell death. Proper apoptosis prevents the development of tumors; however, this mechanism also declines with age, thereby increasing the risk of cancer over time.

10.2.8 Constant Metabolic Potential Theory

CR was noticed in the early 1900s to increase the lifespan of laboratory animals. It became clear that animals with high metabolic rates lived shorter lives. Mice with a resting heart rate of over 300 live only for 2 years on average, cats for 15 years, while elephants with a heart rate of 25 might live for 60 years. The human body has a moderate metabolic rate that is associated with a longer lifespan. One of the effects of CR is that it reduces the activity of the IGF-1 pathway, which may play a role in longevity. A higher level of IGF-1 decreased the lifespan of laboratory animals. In addition, CR increases the activity of antioxidant enzymes, thereby reducing cellular oxidative damage. It is important to note that the higher basic metabolic rate decreases lifespan, rather than the temporary increase gained upon physical training. For example, Tour de France athletes consume over 8000 kcal daily during the race, and they live longer compared to the untrained general population (Sanchis-Gomar et al., 2011).

10.2.9 Neuroendocrine Control Theory

This theory emphasizes the decreased sensitivity of receptors in the central nervous system (mainly hypothalamic) and peripheral hormone receptors. It influences significantly the amount of hormones, which regulate stimulating and inhibitory processes. Alteration in hormone levels influences several cellular functions, which finally promotes aging processes according to this theory. An example based on this neuroendocrine theory by Dean and Dilman is insulin resistance, where glucose uptake from the blood is inefficient even with a high level of circulating insulin. Dilman emphasized the role of declined sensitivity of adrenaline, noradrenaline, and dopamine receptors in aging.

10.2.10 Impaired Immune System Theory

Incidence of several diseases increases with age, which is thought to be the consequence of impaired functions of the immune system. Indeed, the amount and activity of the immune cells decline with age; for instance, the size of the thymus, which produces immune cells, decreases significantly after a certain age. One consequence of impaired immune functions is the improper distinction between a harmful substance and an individual's own bodily cells, and therefore their own tissue provokes an immune response. This process is called autoimmune response, by which the immune cells attack and damage bodily tissues by mistake. Defense against bacterial and viral pathogens may weaken. This theory, established by Walford, explains many aspects of aging.

10.3 PHYSICAL TRAINING AND FREE RADICAL THEORY OF AGING

Free radical theory is the focus of studies that explore the effects of physical training on aging processes, therefore we discuss only this theory. Free radicals are a side product of aerobic metabolism. The estimated amount of free radicals produced daily in the body is about 1 million, and these react shortly after production with proteins, lipids, and DNA. The consequences of these oxidative chain reactions include impaired protein function, alteration of membrane fluidity, and DNA mutations. Oxidative damage of cellular organelles influences cellular functions, organs, and eventually bodily functions. A cause-and-effect relationship has been claimed between accumulation of oxidative damage and aging.

Antioxidant administration decreased oxidative damages in laboratory animals, which resulted in improved brain function measured by mistakes in a labyrinth test (Carney et al., 1991). The level of oxidative damage and brain functions returned to the original aging state upon cessation of antioxidant treatment, implying that oxidative damage causes impaired functions. Our group investigated the effect of physical training instead of antioxidant intake. Exercising prevented aging-related impairment of brain functions measured by memory and learning tests, and it was associated with lower oxidative damage of proteins (Radak et al., 2001). In addition to the brain, oxidative protein and DNA damage and lipid peroxidation were ameliorated or prevented in the liver and skeletal muscle by frequent physical training (Bain et al., 1979; Radak et al., 2004).

The cause behind the preventive effect of physical training is its stimulating effect on the antioxidant system and repairing enzymes (Radak et al., 2008a,b). Our group has demonstrated that frequent physical training increases the activity of DNA repair enzymes in the nucleus, which decreases the amount of DNA damage and number of mutations. It is well-known that incidence of cancer rises with age, and physical training decreases the incidence of breast, prostate, and colon cancer by 50% (Lacza and Radak, 2013; Radak et al., 2008a). According to our hypothesis, decreased DNA damage lies behind the preventive effect of physical training.

Frequent physical training ameliorates oxidative stress during aging due to stimulated activity of antioxidant system and repairs enzymes according to current knowledge, which results in improved functions and decreased incidence of certain disorders.

10.4 PHYSICAL TRAINING AND AGING OF THE CENTRAL NERVOUS SYSTEM

Decline in brain functions occurs after the age of about 60, and the incidence of two main central nervous system disorders increases: Alzheimer's and Parkinson's diseases. Dementia refers to a progressive decline in brain functions, and its two main forms are Alzheimer's and Parkinson's. Risk factors include physical inactivity in both cases. The number of neurons declines with age since they cannot divide, and also the consistence of both white and gray matter is impaired (Colcombe et al., 2004b). Brain white and gray matter of elderly individuals with better physical fitness showed better consistency, thus physical exercising may inhibit or at least slow the thinning of brain structures during aging (Colcombe et al., 2004a). Elderly individuals with good physical fitness showed better performance in brain functional testing, and more neurons were recruited to solve certain functional tasks demonstrated by noninvasive functional magnetic resonance imaging (fMRI) (Colcombe et al., 2004a). Untrained elderly individuals were then exposed to 6 months of physical training and tested again, which resulted in better performance in brain functional tests, and recruitment of neurons showed a similar increase. Functional imaging also revealed that the size of white and gray matter was larger in elderly individuals with better aerobic endurance (Colcombe et al., 2006). Furthermore, the size of the hippocampus (the center for memory and learning) and learning efficiency can also be improved by frequent physical training (Erickson et al., 2011). Although these human studies are very important, the exploration of mechanisms behind these phenomena is difficult due to limited sampling. Therefore, animal studies provide important information on the mechanisms. For instance, it has been demonstrated that new neurons emerge from stem cells in the brain upon physical exercise (van Praag et al., 1999b); this might be the cause behind better cognitive functions in trained individuals. In addition, the level of brain derived neurotrophic factor (BDNF) increased following exercise; BDNF is responsible for learning and memory, neurogenesis, and neuronal vitality (Radak et al., 2014). Physical training increases the activity of antioxidants and repairs enzymes in organs including the brain, therefore it provides more efficient protection against oxidative damage due to enhanced free radical production with age. Physical training enhances the degradation of beta-amyloid. Accumulation of this protein in the brain is a sign of Alzheimer's disease, which results in serious functional deterioration. The activity of neprilysin enzymes, which degrades beta-amyloid, was increased by voluntary physical exercise in transgenic mice producing extra amounts of beta-amyloid, therefore the amounts of beta-amyloid were reduced (Radak et al., 2010). This study proves the beneficial effect of physical training in preventing Alzheimer's disease. Similar effects have been shown in Parkinson's disease. Physical exercise also increases the density of the capillary network in the brain, which is an important factor in preventing brain aging.

In summary, brain functions and its structure are affected by aging. Frequent physical training is a well-known natural tool to ameliorate and prevent aging-related neurodegeneration.

10.5 PHYSICAL TRAINING AND CARDIOVASCULAR AGING

Heart functions show significant differences in the young and elderly. They include arteries; their wall thicken with age and become inelastic partly due to increased collagen content. Rigidity of vessels results in higher peripheral resistance associated with higher blood pressure. It should be noted that lipoprotein level influences vascular rigidity since high levels of triglycerides and low-density lipoproteins (LDL), and their free radical-related oxidation and accumulation on the vessel wall result in atherosclerosis, by which the vessel lumen is reduced; this is a risk factor of heart failure. Triglyceride and LDL level increase after the age of 40–50, whereas the level of high-density lipoproteins (HDL) is unchanged. The size of the heart also increases with age, primarily the thickness of the wall between the ventricles. However, this is not associated with better heart functions; rather its performance is impaired. This kind of thickening is pathophysiological.

In contrast, increased muscle mass of the left ventricle in response to physical training improves heart functions. It results in bradycardia (a lower resting heart rate), which improves endurance of the heart. Frequent endurance training increases stroke volume, and reduces age-related functional deterioration of the cardiovascular system. Frequent exercising also reduces the levels of triglyceride and LDL and increases HDL concentration, which protects against atherosclerosis. Endothelial nitrogen oxide synthase (eNOS) plays an extremely important role in regulation of vessel wall elasticity. Increased activity of eNOS due to physical exercising promotes cardiovascular health. Thus, aerobic physical training slows aging-related structural and functional deterioration of the heart (larger but less contractile cells), and it protects against cardiovascular diseases and increases lifespan through increased wall thickness and volume of the left ventricle, improved vegetative innervation, decreased peripheral resistance, and decreased levels of cholesterol.

Maximal oxygen uptake also decreases after 30 years of age. Functional deterioration of the heart and lungs are the main factors in this process, and it also can be slowed down by physical training. A high capacity of oxygen uptake protects against several health disturbances, while low maximal oxygen uptake capacity is a risk factor of lifestyle diseases.

Decreased lung capacity is one of the most important indexes during aging. Alveolar size and surface decrease significantly with age, resulting in decreased lung capacity. In addition, the decreased surface for gas exchange directly impairs efficiency of oxygen uptake. Vascular density of lung decreases with age, and capillaries deteriorate. Age also decreases the elastic components of the lung. The diaphragm, responsible for breathing, reduces in size and strength similar to other skeletal muscles; however, this can be prevented by physical training through strengthening the obturator muscle and maintaining elasticity of pleura.

10.6 PHYSICAL TRAINING AND AGE-RELATED ALTERATIONS OF BONES AND MUSCLES

Osteopenia/osteoporosis or bone thinning is one of the most frequent forms of age-related deterioration; it is influenced by estrogen, vitamin D, intestinal Ca^{2+} absorption, and physical inactivity. Lower estrogen levels due to menopause cause impaired intestinal Ca^{2+} absorption, thus the Ca^{2+} demand of the body is fulfilled by mobilizing calcium from the bones. Low vitamin D levels result in similar consequences. Skin contains less pro-D with age, which influences calcium absorption in the intestines. Thus, bone thinning is a complex phenomenon, and affects mainly women due to their lower bone mass and density. Bones are renewing continuously similar to other tissues; in this process, cells that break down (osteoclast) and build (osteoblast) bone tissue play the major role. The efficiency of building processes declines with age, resulting in decreased bone density. Menopause or surgical removal of ovaries leads to decreased estrogen levels, putting these individuals at a higher risk of osteoporosis, which increases the risk of bone fractures. Frequent physical training ameliorates bone thinning since the force generated by muscle contractions stimulates bones to maintain their structure and density.

Joint tissue also deteriorates with age. Cartilage in joints exposed to high loading, such as the hip and knee, decays with age, especially in obese and overweight individuals. Decreased synovial fluid in the joint cavity increases the risk of joint injuries due to higher viscosity, and this is more frequent in the elderly. Physical training generally improves joint wellness; however, high-effort sports such as athletic throwing, ball games, and endurance sports (e.g., walking) may result in joint issues.

Skeletal muscle constitutes about 40% of total body weight in the young, which also deteriorates with age, resulting in a 40%–50% loss in elderly and impaired function. Physical activity drops over the age of 60, and untrained muscles lose their weight. Skeletal muscle is specialized in locomotion; however, it also plays a role in regulating blood sugar level. Although type 2 diabetes is mainly a lifestyle disease, age also has an influence. Lifestyle and age-related hyperglycemia (insulin resistance) is a more common phenomenon in the modern world, and associated with an inactive lifestyle. It correlates with loss of muscle mass. The GLUT4 glucose transporter is responsible for the uptake of glucose from blood in skeletal muscle, where it is converted into glycogen and used as an energy source or converted to fat. Physical activity demands a high amount of energy, thus it stimulates the muscle to consume its glucose content and provide ATP for muscle contractions; alternatively, it stores glucose for later use in the form of glycogen, easily available, rather than in the form of fat tissue. Frequent physical training increases the amount of GLUT4 in skeletal muscle, resulting in a higher capacity for glucose uptake, which prevents the insulin resistance that leads to diabetes. Skeletal muscle mass and the amount of GLUT4 receptors decline with age, which may predispose the elderly to develop diabetes. This can be prevented by physical training. Individuals suffering from type 2 diabetes also benefit from physical exercise, which ameliorates the severity of diabetes.

Muscle fiber composition also changes with age; the amount of type II muscle fibers decline with age, which can be explained by the decreased amount of mitochondria due to degeneration. The mass of contractile components declines after the age of 60; this is called sarcopenia and seems to be associated with inflammation. The level of TNF-α pro-inflammatory cytokine increases with age, and this molecule induces nuclear transcription factor NF-κB, which regulates inflammatory processes. Myostatin is also activated in this process, promoting a further decrease in muscle mass.

Aging of skeletal muscle can be prevented effectively by frequent physical training; this should be comprised of resistance training, which induces muscle growth, and endurance training. According to a recent investigation, aerobic training is able to normalize the balance of catabolic and anabolic processes in skeletal muscle of aged mice. Aerobic training also inhibits age-related increase of myostatin and decrease of follistatin (Ziaaldini et al., 2015), thus physical training prevented the intense breakdown of proteins, which is the main cause of muscle loss in the elderly.

10.7 "R" EXTRA

Although the free radicals theory of aging is the most commonly accepted concept, antioxidant intake does not result in longer lifespans in higher organisms. Free radicals in lower amounts are important signaling molecules in cellular functions; however, prolonged exposure in high concentrations promotes oxidative damage to DNA, proteins, and lipids. DNA damage is especially dangerous since it may result in cancer. Metabolism correlates with the production of free radicals, which is regulated by sirtuins. Since sirtuins are NAD-dependent enzymes, the NAD/NADH ratio influences the activity of sirtuins. Sirtuins, mainly SIRT1, regulate the activity of many transcription factors that determine oxidant and antioxidant processes (Radak et al., 2013). This relationship shows that the free radical theory and sirtuin theory of aging are interconnected. One beneficial effect of frequent physical training is that it increases the activity of housekeeping enzymes, which repair or eliminate damaged proteins. Sirtuins are also involved in this process.

10.8 SUMMARY

Adaptation of human body to external challenges declines with age, which results in impaired homeostasis. Physical inactivity further impairs adaptive responses and promotes the development of lifestyle diseases, type 2 diabetes, cardiovascular diseases, Alzheimer's and Parkinson's diseases, osteoporosis, and age-related muscle atrophy. Frequent physical training is very important for the elderly. Average physical activity of young individuals is higher than that of the elderly, thus negative effects of physical inactivity are more significant in the latter, and therefore they are recommended to make time for frequent physical exercise. It provides an opportunity to maintain balance and prevent lifestyle diseases, and improve life quality and increase lifespan.

TEST QUESTIONS

1. What are the most important aging theories?
2. What is the maximal lifespan of humans and what factors determine average life expectancy?
3. What is a telomere and what is its role in aging?
4. What are free radicals and what is their role in aging?
5. What are the characteristics of age-related changes of the heart and how are they influenced by physical training?
6. How does physical training influence the aging of the nervous system?
7. How does physical training influence bone thinning?

BIBLIOGRAPHY

Bain, J., Duthie, M., Keene, J., 1979. Relationship of seminal plasma testosterone and dihydrotestosterone to sperm count and motility in man. Arch. Androl. 2, 35–39.
Carney, J.M., Starke-Reed, P.E., Oliver, C.N., Landum, R.W., Cheng, M.S., Wu, J.F., Floyd, R.A., 1991. Reversal of age-related increase in brain protein oxidation, decrease in enzyme activity, and loss in temporal and spatial memory by chronic administration of the spin-trapping compound N-tert-butyl-alpha-phenylnitrone. Proc. Natl. Acad. Sci. U. S. A. 88, 3633–3636.
Colcombe, S.J., Erickson, K.I., Raz, N., Webb, A.G., Cohen, N.J., McAuley, E., Kramer, A.F., 2003. Aerobic fitness reduces brain tissue loss in aging humans. J. Gerontol. A Biol. Sci. Med. Sci. 58, 176–180.
Colcombe, S.J., Kramer, A.F., Erickson, K.I., Scalf, P., McAuley, E., Cohen, N.J., Webb, A., Jerome, G.J., Marquez, D.X., Elavsky, S., 2004a. Cardiovascular fitness, cortical plasticity, and aging. Proc. Natl. Acad. Sci. U. S. A. 101, 3316–3321.
Colcombe, S.J., Kramer, A.F., McAuley, E., Erickson, K.I., Scalf, P., 2004b. Neurocognitive aging and cardiovascular fitness: recent findings and future directions. J. Mol. Neurosci. 24, 9–14.
Colcombe, S.J., Erickson, K.I., Scalf, P.E., Kim, J.S., Prakash, R., McAuley, E., Elavsky, S., Marquez, D.X., Hu, L., Kramer, A.F., 2006. Aerobic exercise training increases brain volume in aging humans. J. Gerontol. A Biol. Sci. Med. Sci. 61, 1166–1170.
Erickson, K.I., Voss, M.W., Prakash, R.S., Basak, C., Szabo, A., Chaddock, L., Kim, J.S., Heo, S., Alves, H., White, S.M., Wojcicki, T.R., Mailey, E., Vieira, V.J., Martin, S.A., Pence, B.D., Woods, J.A., McAuley, E., Kramer, A.F., 2011. Exercise training increases size of hippocampus and improves memory. Proc. Natl. Acad. Sci. U. S. A. 108, 3017–3022.

Harman, D., 1956. Aging: a theory based on free radical and radiation chemistry. J. Gerontol. 11, 298–300.

Lacza, G., Radak, Z., 2013. Is physical activity an elixir? Orv. Hetil. 154, 764–768.

Lazarov, O., Robinson, J., Tang, Y.P., Hairston, I.S., Korade-Mirnics, Z., Lee, V.M., Hersh, L.B., Sapolsky, R.M., Mirnics, K., Sisodia, S.S., 2005. Environmental enrichment reduces Abeta levels and amyloid deposition in transgenic mice. Cell 120, 701–713.

Linskens, M.H., Harley, C.B., West, M.D., Campisi, J., Hayflick, L., 1995. Replicative senescence and cell death. Science 267, 17.

Radak, Z. (Ed.), 2004. Exercise and Diseases: Prevention Through Training. Meyer and Meyer Sport, Aachen.

Radak, Z., Kaneko, T., Tahara, S., Nakamoto, H., Ohno, H., Sasvari, M., Nyakas, C., Goto, S., 1999. The effect of exercise training on oxidative damage of lipids, proteins, and DNA in rat skeletal muscle: evidence for beneficial outcomes. Free Radic. Biol. Med. 27, 69–74.

Radak, Z., Kaneko, T., Tahara, S., Nakamoto, H., Pucsok, J., Sasvari, M., Nyakas, C., Goto, S., 2001. Regular exercise improves cognitive function and decreases oxidative damage in rat brain. Neurochem. Int. 38, 17–23.

Radak, Z., Naito, H., Kaneko, T., Tahara, S., Nakamoto, H., Takahashi, R., Cardozo-Pelaez, F., Goto, S., 2002. Exercise training decreases DNA damage and increases DNA repair and resistance against oxidative stress of proteins in aged rat skeletal muscle. Pflugers Arch. - Eur. J. Physiol. 445, 273–278.

Radak, Z., Chung, H.Y., Naito, H., Takahashi, R., Jung, K.J., Kim, H.J., Goto, S., 2004. Age-associated increase in oxidative stress and nuclear factor kappaB activation are attenuated in rat liver by regular exercise. FASEB J. 18, 749–750.

Radak, Z., Toldy, A., Szabo, Z., Siamilis, S., Nyakas, C., Silye, G., Jakus, J., Goto, S., 2006. The effects of training and detraining on memory, neurotrophins and oxidative stress markers in rat brain. Neurochem. Int. 49, 387–392.

Radak, Z., Chung, H.Y., Goto, S., 2008a. Systemic adaptation to oxidative challenge induced by regular exercise. Free Radic. Biol. Med. 44, 153–159.

Radak, Z., Chung, H.Y., Koltai, E., Taylor, A.W., Goto, S., 2008b. Exercise, oxidative stress and hormesis. Ageing Res. Rev. 7, 34–42.

Radak, Z., Hart, N., Sarga, L., Koltai, E., Atalay, M., Ohno, H., Boldogh, I., 2010. Exercise plays a preventive role against Alzheimer's disease. J. Alzheimers Dis. 20, 777–783.

Radak, Z., Koltai, E., Taylor, A.W., Higuchi, M., Kumagai, S., Ohno, H., Goto, S., Boldogh, I., 2013. Redox-regulating sirtuins in aging, caloric restriction, and exercise. Free Radic. Biol. Med. 58, 87–97.

Radak, Z., Ihasz, F., Koltai, E., Goto, S., Taylor, A.W., Boldogh, I., 2014. The redox-associated adaptive response of brain to physical exercise. Free Radic. Res. 48, 84–92.

Sanchis-Gomar, F., Olaso-Gonzalez, G., Corella, D., Gomez-Cabrera, M.C., Vina, J., 2011. Increased average longevity among the "Tour de France" cyclists. Int. J. Sports Med. 32, 644–647.

Szilard, L., 1959. On the nature of the aging process. Proc. Natl. Acad. Sci. U. S. A. 45, 30–45.

van Praag, H., Christie, B.R., Sejnowski, T.J., Gage, F.H., 1999a. Running enhances neurogenesis, learning, and long-term potentiation in mice. Proc. Natl. Acad. Sci. U. S. A. 96, 13427–13431.

van Praag, H., Kempermann, G., Gage, F.H., 1999b. Running increases cell proliferation and neurogenesis in the adult mouse dentate gyrus. Nat. Neurosci. 2, 266–270.

Ziaaldini, M.M., Koltai, E., Csende, Z., Goto, S., Boldogh, I., Taylor, A.W., Radak, Z., 2015. Exercise training increases anabolic and attenuates catabolic and apoptotic processes in aged skeletal muscle of male rats. Exp. Gerontol. 67, 9–14.

CHAPTER 11

Sport Genetics

Over the past few decades, the role that genetics plays in sport performance has verified many of the beliefs of heritability put forth by scientists before techniques were available to prove their visual observations. For example, early work with winning racehorses resulted in highly successful offspring when the animals were crossbred (Schroder et al., 2011). Geographical correlations have also been noted, such as the high success rate of east African athletes in middle and long-distance running, and successful sprint running and long jumping, which requires the application of a large amount of force, by West Africans, African Americans, and Caribbean athletes with a heritage from West Africa. These observations emphasize the role that genetic factors can play in professional sport (Rankinen et al., 2010). Knowledge of the genetic factors may have an important role in sport category selection and in individual training. Genes are regions (locus) of DNA encoding protein products, and are the molecular unit of heredity. Recently a great deal of effort has been employed to identify genes that are responsible for excellent sport performance or the development of sport skills. Several genes have been associated with physical performance; however, this complex issue is complicated by the interference of genetic factors, pseudogenes, or micro-RNAs. Furthermore, genetic information can be affected by reversible epigenetic modification of DNA by histone proteins, which can be affected by lifestyle.

Since sport performance is affected by many factors such as physiological, psychological, environmental, technical, and tactical, the presence of a gene itself, and its activation or repression, may seem to have subtle effects; however, it may affect sport performance. The importance and effects of genetic discoveries will be extremely important for sport in the future.

Skeletal musculature is important for physical training, and it is one of the most plastic organs of the human body. It can adapt easily, its size can be developed easily, its coordination with other motor units can be improved easily, and significant changes can be attained in its metabolic condition. The underlying causes of these characteristics of the skeletal musculature are induced by genetic and epigenetic mechanisms, which will be addressed in this chapter.

 171

11.1 EPIGENETIC MODIFICATIONS

Epigenetic variations are structural modifications in the chromosomes without changes in the DNA nucleotide sequence. These modifications can be caused by environmental and external factors such as lifestyle, diet, and physical training. The mechanisms of epigenetic modification comprise DNA methylation and methylation, acetylation, ubiquitination, and carbonylation of histone proteins in the chromatin (Fig. 11.1).

DNA is stored in the nucleus of every cell. Because of its very large size, DNA is organized in a complex of DNA and proteins called chromatin. The DNA is comprised of 146 base pairs, each of which wraps around histone proteins forming nucleosomes, and histones are able to regulate the binding of transcriptional factors to DNA. Gene transcription, by which a particular segment of DNA is copied into mRNA, is initiated by the binding of transcriptional factors. This process can only

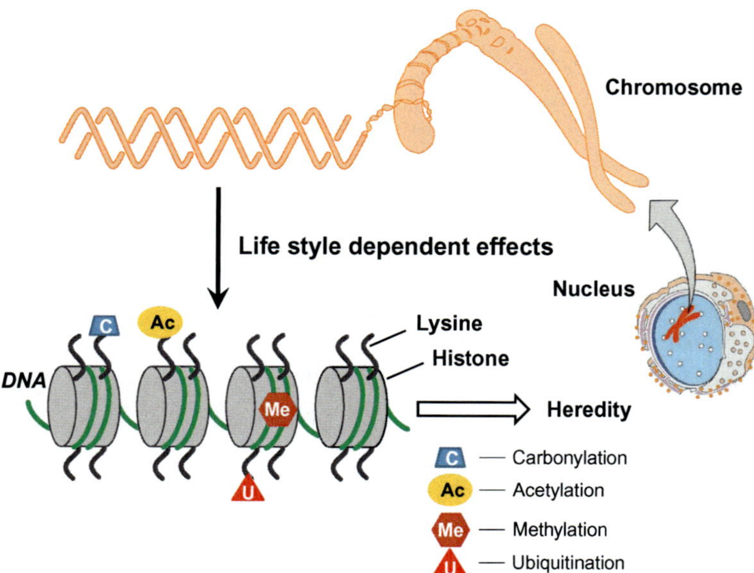

Fig. 11.1 Epigenetic modifications. Epigenetic variations are reversible and inheritable changes in the DNA structure and histone proteins found in the nucleus. These modifications switch genes on and off without causing changes in the DNA nucleotide sequence. Methylation, acetylation, carbonylation, or ubiquitination are the main mechanisms that are applied to regulate genes. Lifestyle has an effect on epigenetic modifications, and therefore physical training and diet may play major roles in epigenetic phenotypic trait variations. This implies that we are able to influence not just our own quality of life but that of our descendants as well.

be executed if the association between DNA and histone proteins is weakened in a given time and place.

The strength of the attachment between the DNA and histones depends on the quality and quantity of the posttranslational modifications. For instance, the acetylation of histone proteins weakens the DNA-histone conjunction, increasing the availability of a particular DNA segment for activation, whereas deacetylation has the opposite effect, causing the silencing of genes.

The acetyl group carries an electric charge, and therefore it can influence the interaction between the DNA and histone proteins. Acetylation is regulated by histone acetyl-transferases (HAT), whereas deacetylation is executed by histone-deacetylases (HDAC). The latter are able to modify the lysine amino acid in any proteins, not only in histones. Phosphorylation of HAT enzyme regulates its activity, allowing the alteration of histone-DNA attachment, causing a gene to be switched on.

Regular physical training has a long-term impact on the acetylation of a particular part of histone 3 (H3) proteins in the hippocampus causing increased transcription of the BDNF gene (brain derived neurotrophic factor) (Gomez-Pinilla et al., 2011). This epigenetic modification of the BDNF gene results in improved memory and alleviation of symptoms of depression (see in Chapter 9 about prevention). Long-term changes in histone acetylation can cause long-term adaptation in epigenetic regulation. Histone proteins can also be methylated; single or multiple methylations of lysine and arginine amino acids are common. Although attachment of a methyl group on the amino acid does not alter its electric charge, methylation of certain amino acids (e.g., lysine in the ninth position in H3K9 histone 3) silences, and then inhibits, the transcription of certain genes.

Another example of epigenetic phenotypic alteration is the change in hair color and lifespan of laboratory mice caused by the methylation of the Agouti gene (Dolinoy et al., 2010).

The first epigenetic observation in the case of skeletal muscle was the detection of the differential physiological characteristics of the cells isolated from slow- and fast-twitch muscle fibers caused by the differences in the light and heavy chains of myosin (Matsuda et al., 1983). It is not surprising that cells isolated from slow-twitch muscle fibers contain distinctive myosin heavy chains (MHC I) peculiar to that fiber, which enable long endurance feats and their shape differs from that isolated from fast-twitch muscle fibers (MHC Iib). However, it is surprising that these differences are caused by epigenetic modifications.

Slow-twitch muscle fibers are innervated by 5–10 Hz nerves, and therefore MHC I synthesis is promoted by this type of innervation. It has been proven that there is a cause-and-effect relationship in the transcription of the different myosin heavy chains in the slow-twitch and fast-twitch muscle fibers, and the acetylation and trimethylation of H3 histones (Pandorf et al., 2009). This observation implies that the emergence of slow-twitch and fast-twitch muscle fibers are altered by epigenetic factors, which raises the opportunity to alter the ratio between the different types of muscle fibers.

Bicycle ergometer training at 75% of VO_2max has been shown to cause increased acetylation of H3K36 without any changes being seen in the rate of acetylation of H3K9 and H3K14 in the skeletal muscle of young males (McGee et al., 2009). Changes in the enzymatic activity of AMPK and calcium-calmodulin dependent protein kinase (CaMKII) have also been observed, and this regulates the activity of class II histone deacetylase (IIa HDAC). These observations suggest that adaptation following physical training may cause complex and epigenetic modifications.

The musculature of highly trained athletes contains significant amounts of glycogen, 50%–100% more than untrained muscle, which is the result of epigenetic changes caused by adaptation.

For example, sirtuins deacetylate peroxisome proliferator activated receptor-γ coactivator-1α (PGC-1 alfa), resulting in significant increases in the number of mitochondria and an increased ratio of slow-twitch muscle fibers.

Lifestyle and obesity have an effect on the musculature through epigenetic alterations. Although there are no sequential changes in the DNA, the transcription of stearoyl CoA desaturase gene increases in obese individuals, causing increased production of myostatin (Hittel et al., 2009), which leads to decreased muscle weight and increased fat content in the musculature.

Age also has an effect on epigenetic changes in the nucleosomes, since observations in young and elderly twins show differences in DNA methylation, which may have a role in the decrease of musculature in the elderly (Fraga et al., 2005).

Since environmental and external factors cause epigenetic changes, the determination of the degree of their inheritability has significant importance. Experiments with laboratory mice have shown that certain epigenetic changes are heritable (Wisloff et al., 2005). If this finding applies to human populations, it raises the concern of the responsibility of the parents since their lifestyle would have a major effect on the health and sport abilities of their children.

11.2 GENETICS AND SPORT

The gene coding angiotensin converting enzyme (ACE) is one of the most researched genes in sport genetics. Several variants of the ACE gene have been claimed to be associated with fitness and endurance. Athletes with decreased ACE circulating in the blood demonstrate excellent performance in endurance events. ACE is an enzyme that converts angiotensin I to angiotensin II. The latter plays a role in cell division, vascularization, Na^+ absorption, and stress reactions. The renin-angiotensin-aldosterone system is an important regulator of blood pressure and volume homeostasis. ACE also promotes the metabolism of bradykinin; it has a membrane-bound and a circulating free isoform. Bradykinin also has a vasodilator effect, and therefore it decreases blood pressure. Accordingly, ACE inhibitors decrease blood pressure and inhibit the metabolism of bradykinin.

The gene encoding ACE is in chromosome 17. Different alleles of the ACE gene are associated with aerobic endurance. An allele is a variant form of a gene. Some genes have a variety of different forms, which are located at the same position or genetic locus, on a chromosome. Somatic cells are diploids because they have two alleles at each genetic locus with one allele inherited from each parent: a total of 23 from the father and 23 from the mother. This means that there are two copies from each gene in the cell. In the case of two identical copies the gene are termed homozygous, and if the two copies are different the term is heterozygous. The effect on phenotype of one allele is the ability to mask the contribution of a second allele of the same locus, i.e., the first allele is dominant and the second allele is recessive. For example, development of eye color is a result of the conflict of the rival parental alleles (Fig. 11.2).

A single nucleotide polymorphism (SNP or snip) is a variation of a single nucleotide in DNA. For example, at a specific base position, the base cytosine may appear in some individuals, but the position is occupied by base thymine in others. SNPs are not just associated with genes; they can also occur in non-coding regions of DNA. Base variations are common in DNA as every 300th base differs; however, some variations are present to some appreciable degree within a population (>1%), the latter being called SNP (Fig. 11.3). Generally, SNPs do not affect function; however, there are some variations in genes that cause an individual to be predisposed to develop a certain disease.

No definite correlation has been found between ACE in/del alleles and the outstanding performance of athletes from Kenya and Ethiopia (Ash et al., 2011; Scott et al., 2005). However, there are some observations that suggest the ACE gene does affect performance in response to training (Vaughan et al., 2013).

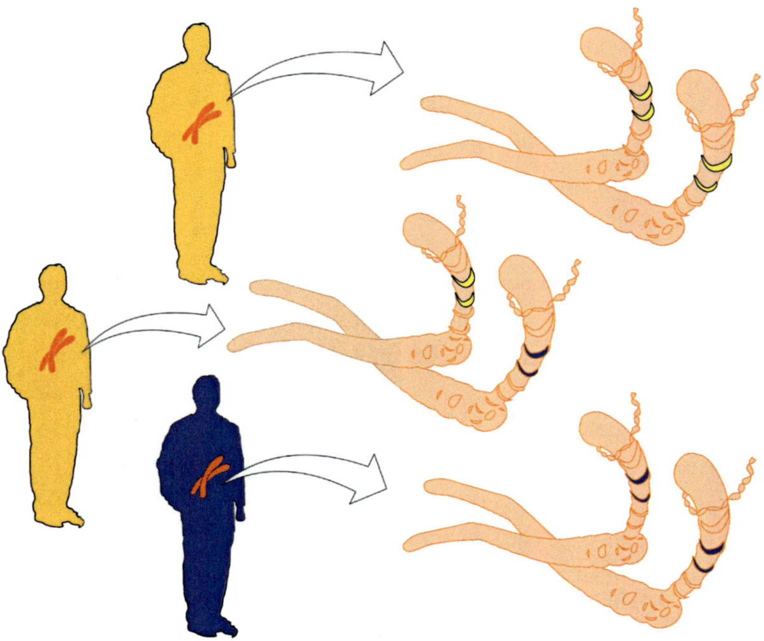

Fig. 11.2 Allele. An allele is a variant form of a gene, a difference in DNA sequence.

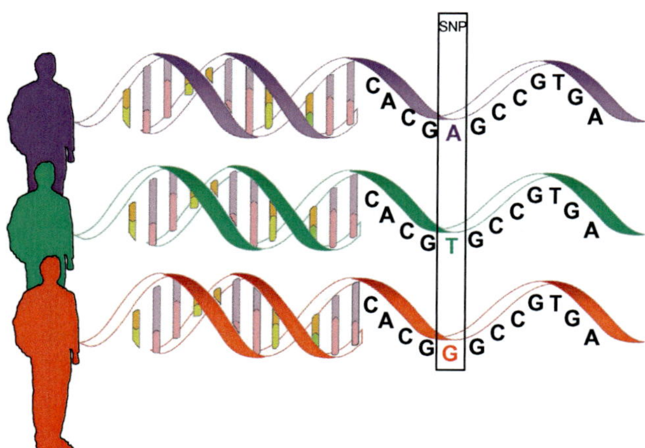

Fig. 11.3 Single nucleotide polymorphism (SNP). The ACE-insertion allele (ACE-in) appears frequently in athletes who compete in sport events that require aerobic endurance (Alvarez et al., 2000; Sonna et al., 2001). However, other observations seem to prove the opposite (Day et al., 2007). The ACE-deletion allele (ACE-del) appears in power athletes requiring strength. However, the relationship between ACE genotype and metabolic efficiency needs to be demonstrated. In the case of ACE-in, there is an extra nucleotide in the ACE allele, whereas in the case of ACE-del, a nucleotide base pair is missing in the allele.

Regardless of the contradictory observations for individual differences in endurance, strength, age, and ethnic group, ACE-in is believed to bring about an advantage in aerobic endurance sports, whereas ACE-del is related to advantage in sports that require strength.

Definite physiological and biochemical characteristics of ACE-in/del alleles are not known; however, there is an indication of ACE-in allele in improved performance in response to training, especially for the cardiovascular system.

Allele variations in ACE gene or other genes may be important in certain skills, either as advantages or disadvantages. ACE gene is associated with bradykinin receptor B2, which can activate MAPK; therefore, the bradykinin receptor may play a role in inside cell homeostasis. It is debated whether the polymorphism of bradykinin receptor B2 is associated with sport performance (Grenda et al., 2014).

AGTR2 is a receptor for angiotensin II. The presence of the C allele in *AGTR2* gene shows a significant correlation with the number of slow-twitch muscle fibers and performance based on measurements on Polish endurance athletes, whereas the A allele of the gene shows a correlation with the number of fast-twitch muscle fibers and performance in power athletes (Mustafina et al., 2014).

Adenosine monophosphate deaminase (AMPD) is one of the key enzymes in the metabolic cycle of the cell, and has significant importance in endurance training. Mutation of the AMPD gene has been shown in outstanding endurance athletes, suggesting its role in sport performance (Rubio et al., 2005).

The dominance of East African runners in middle and long-distance events indicates the role of genetic factors in sport performance. The most successful Kenyan runners are from two ethnic groups, Kalenjin and Nandi from the Rift Valley, which again indicates the relationship between genetic factors in performance and endurance. However, the fact that these runners are in training from childhood emphasizes the importance of lifestyle as well (Onywera et al., 2006).

Several analyses have been carried out to explore the role of genes in the differences in the performance of Caucasian and African long-distance athletes. However, the results are still controversial. Although the long-distance runners from Africa have smaller body mass, there is no difference in VO_2max when compared to European elite runners. However, it seems that African runners have better running efficiency and better heat regulation (Marino et al., 2004). Moreover, it has been also reported that East

African runners have a lower percentage of slow and higher percentage of fast-twitch oxidative mixed fibers compared to Caucasian runners, while the activity of those enzymes that are important oxidative metabolism is very comparable (Kohn et al., 2007).

Biogenesis and plasticity of mitochondria may be of great importance in sports that require endurance since more mitochondria are able to produce more ATP in a given time frame. Mitochondria also play a role in programmed cell death, mitochondria-related diseases, and the adaptation processes during physical training. Mitochondrial DNA (mtDNA) decodes only 13 proteins, all of which play a role in oxidative phosphorylation and are essential in sustaining life. The important role of mitochondria in adaptation with physical training raise the questions whether there are genetic differences in mtDNA in trained and untrained athletes, and whether there are correlations with sport categories. The C allele of *PPARD* C294T was found to correlate with athletes of different VO_2max levels (Akhmetov et al., 2007). In the *PPARGC1A* gene, the allele Ser482 was significantly lower in Spanish athletes and cyclists with high VO2max than in the average population (Lucia et al., 2005). However, there are some opposite observations related to the *PPARD* and *PPARGC1A* genes that do not support this notion.

Bouchard's group has provided a great deal of data concerning the relationship between genetic factors and physical sport performance. This group explored the role of mitochondrial genes and showed correlations between mtDNA sequence variations and maximal oxygen uptake and endurance training (Dionne et al., 1993). Two Finnish authors examined the differences between long-distance and short-distance runners in relation to the number of haplogroups of mtDNA. Haplogroups define differences in mtDNA, on an arbitrary A–Z list; they also indicate the origin of a population. K and J groups indicate Middle Eastern origin. According to the Finnish authors, the short-distance runners more frequently possessed K and J haplogroups compared to long-distance running athletes (Niemi and Majamaa, 2005). This would indicate that outstanding sprinters are likely to be from Middle Eastern countries. However, for a variety of reasons, this fact has not been demonstrated.

Optimal genetic maps of mitochondrial biogenesis have been created for endurance athletes, and these have been used for further studies to examine untrained individuals and power athletes. Unfortunately, the model has proven to be unsuccessful because of the strong masking effects of individual differences (Eynon et al., 2011). Nonetheless, Marton's group has shown a correlation between mitochondrial biogenesis proteins and aerobic endurance training (Marton et al., 2015), which indicates that genes involved in mitochondrial biogenesis might be important in endurance training.

A strong correlation between the polymorphism of the HIF1 alfa gene and endurance has been observed as HIF1 alfa influences oxygen transport, vascularization, and carbohydrate metabolism. However, others have observed only a weak correlation between HIF1 alfa polymorphism and endurance (Doring et al., 2010).

The thickness of the Z-disc shows differences in slow-twitch and fast-twitch muscle fibers. Slow-twitch fibers have wider and stronger Z-discs, which makes slow fibers more resistant to mechanical stress and less sensitive to muscle hypertrophy. Besides the role of Z-disc in resistance to mechanical stress, it hosts an important protein: alpha-actinin. In the case of power athletes, alpha-actinin-3 (*ACTN-3*) polymorphism was examined and has been demonstrated to play a role in the formation of the two different types of muscle fibers.

ACTN-3 protein is missing from 15% to 17% of the population, and therefore these individuals have disadvantages in power sports. It is interesting that the *ACTN-3* null genotype, which means that it is not present in an individual, shows decreased activity of the glycogen phosphorylase enzyme, resulting in intensified aerobic metabolic activity (Berman and North, 2010). A correlation between *ACTN-3* polymorphism and the performance of outstanding sprinters from Jamaica and the United States has not been found.

For a complex sports such as rhythmic gymnastics, the role of gene polymorphism affecting body composition (*ADRB2, FTO*), joint mobility (*COL5A1*), power (*ACTN3*), and endurance (*ACE)* might be of special importance. A correlation between gene polymorphism affecting body composition, joint mobility, and performance, but not *ACTN3* and *ACE* genes and performance, have been found in one study (Tringali et al., 2014).

In summary, testing genes might help in sport selection for an individual. However, it should be emphasized that sport performance is a complex process and does not depend solely upon genetic factors (Lucia et al., 2010).

11.3 "R" EXTRA

More than 100 genetic variations have been claimed to be associated with professional sport. However, it seems that no special or singular genotypes correlate with sport performance (Bray et al., 2009). Presumably sport performance is affected by one or more specific genes, in all likelihood in coordination with many other factors.

Micro-RNAs (miRNA) are a small noncoding RNA molecule containing about 22 nucleotides that function in RNA silencing and posttranscriptional regulation of gene expression (Fig. 11.4).

DNA in the nucleus encodes approximately 1000 micro-RNA, which can bind to target mRNAs after activation outside the nucleus, resulting in a decrease or cessation of the production of the target protein. Micro-RNAs are also able to cut mRNA, resulting in cessation of the synthesis of the target protein. This finding explains why there are sometimes differences between the levels of mRNA and its protein products. These small miRNAs play a role in the development of certain diseases or phenotypes (Gallagher et al., 2010). That is to say, they play a role in the regulation of metabolic processes and muscle function, and they may have importance in epigenetics and sport performance. It has been shown that miRNAs play a role in endurance and power sport performance. Following endurance training, *RUNX1*, *SOX9*, and *PAX3* targeting miRNAs levels were significantly lower in young males, who responded well to aerobic training compared to those with a weaker response (Keller et al., 2011). These results correlated strongly with

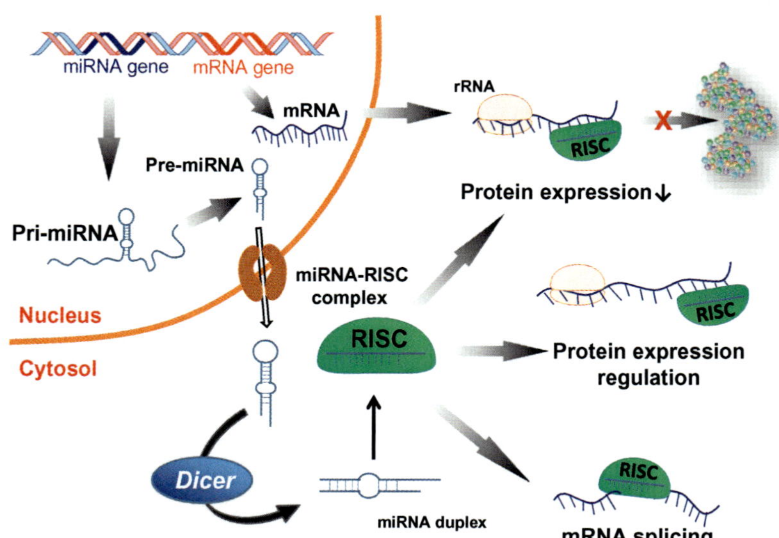

Fig. 11.4 Micro-RNA (miRNA). Binding of miRNA to mRNA is able to regulate protein synthesis. Micro-RNA is activated by an enzyme called Dicer after the progenitor of miRNA is synthesized and departed from the nucleus. The mature miRNA is part of an active RNA-induced silencing complex (RISC) facilitated by Dicer.

those measured in experimental mice, and verify the results measured by Marton's group (Marton et al., 2015).

Similar examinations were done on power performance and miRNA systems were tested on groups that responded well or poorly to power training. Decreased levels of miRNA-378, miRNA-29a, and miRNA-26a were measured in the muscle of the group with the weaker response to training, and the level of miRNA-378 correlated strongly with the increasing size of the muscle cross-sectional area (Davidsen et al., 2011). There are significant individual differences in the level of miRNAs; however, they should be further examined to discover their role in sport performance, as that would be helpful in individual sport selection.

11.4 SUMMARY

Genetics play a very important role in sport performance. However, outstanding performance and individual competence depend on a complex system of abilities of which genes are only a portion.

The complexity of sport performance is reflected by the strong differences in individuals with differential congenital and acquired traits and personality. Epigenetic effects are extremely important, since lifestyle affect not only the life of the individual but its inheritable epigenetic changes reappear in descendants. The positive effects of physical training in prevention and healing manifest through epigenetic modification. The role of genetic factors in professional sports is important; however, a correlation between gene polymorphisms and sport performance has not yet been established.

TEST QUESTIONS

1. What is a gene?
2. What traits can be determined by genetic factors?
3. What is the role of sport genomics?
4. What genes are associated with endurance?
5. What genes are associated with power?
6. What are the advantages and limits of sport genetics?

BIBLIOGRAPHY

Akhmetov, I.I., Astranenkova, I.V., Rogozkin, V.A., 2007. Association of PPARD gene polymorphism with human physical performance. Mol. Biol. 41, 852–857.
Alvarez, R., Terrados, N., Ortolano, R., Iglesias-Cubero, G., Reguero, J.R., Batalla, A., Cortina, A., Fernandez-Garcia, B., Rodriguez, C., Braga, S., Alvarez, V., Coto, E., 2000. Genetic variation in the renin-angiotensin system and athletic performance. Eur. J. Appl. Physiol. 82, 117–120.

Ash, G.I., Scott, R.A., Deason, M., Dawson, T.A., Wolde, B., Bekele, Z., Teka, S., Pitsiladis, Y.P., 2011. No association between ACE gene variation and endurance athlete status in Ethiopians. Med. Sci. Sports Exerc. 43, 590–597.

Berman, Y., North, K.N., 2010. A gene for speed: the emerging role of alpha-actinin-3 in muscle metabolism. Physiology 25, 250–259.

Bray, M.S., Hagberg, J.M., Perusse, L., Rankinen, T., Roth, S.M., Wolfarth, B., Bouchard, C., 2009. The human gene map for performance and health-related fitness phenotypes: the 2006–2007 update. Med. Sci. Sports Exerc. 41, 35–73.

Davidsen, P.K., Gallagher, I.J., Hartman, J.W., Tarnopolsky, M.A., Dela, F., Helge, J.W., Timmons, J.A., Phillips, S.M., 2011. High responders to resistance exercise training demonstrate differential regulation of skeletal muscle microRNA expression. J. Appl. Physiol. 110, 309–317.

Day, S.H., Gohlke, P., Dhamrait, S.S., Williams, A.G., 2007. No correlation between circulating ACE activity and VO2max or mechanical efficiency in women. Eur. J. Appl. Physiol. 99, 11–18.

Dionne, F.T., Turcotte, L., Thibault, M.C., Boulay, M.R., Skinner, J.S., Bouchard, C., 1993. Mitochondrial DNA sequence polymorphism, VO2max, and response to endurance training. Med. Sci. Sports Exerc. 25, 766–774.

Dolinoy, D.C., Weinhouse, C., Jones, T.R., Rozek, L.S., Jirtle, R.L., 2010. Variable histone modifications at the A(vy) metastable epiallele. Epigenetics 5, 637–644.

Doring, F., Onur, S., Fischer, A., Boulay, M.R., Perusse, L., Rankinen, T., Rauramaa, R., Wolfarth, B., Bouchard, C., 2010. A common haplotype and the Pro582Ser polymorphism of the hypoxia-inducible factor-1alpha (HIF1A) gene in elite endurance athletes. J. Appl. Physiol. 108, 1497–1500.

Eynon, N., Ruiz, J.R., Meckel, Y., Moran, M., Lucia, A., 2011. Mitochondrial biogenesis related endurance genotype score and sports performance in athletes. Mitochondrion 11, 64–69.

Fraga, M.F., Ballestar, E., Paz, M.F., Ropero, S., Setien, F., Ballestar, M.L., Heine-Suner, D., Cigudosa, J.C., Urioste, M., Benitez, J., Boix-Chornet, M., Sanchez-Aguilera, A., Ling, C., Carlsson, E., Poulsen, P., Vaag, A., Stephan, Z., Spector, T.D., Wu, Y.Z., Plass, C., Esteller, M., 2005. Epigenetic differences arise during the lifetime of monozygotic twins. Proc. Natl. Acad. Sci. U. S. A. 102, 10604–10609.

Gallagher, I.J., Scheele, C., Keller, P., Nielsen, A.R., Remenyi, J., Fischer, C.P., Roder, K., Babraj, J., Wahlestedt, C., Hutvagner, G., Pedersen, B.K., Timmons, J.A., 2010. Integration of microRNA changes in vivo identifies novel molecular features of muscle insulin resistance in type 2 diabetes. Genome Med. 2, 9.

Gomez-Pinilla, F., Zhuang, Y., Feng, J., Ying, Z., Fan, G., 2011. Exercise impacts brain-derived neurotrophic factor plasticity by engaging mechanisms of epigenetic regulation. Eur. J. Neurosci. 33, 383–390.

Grenda, A., Leonska-Duniec, A., Cieszczyk, P., Zmijewski, P., 2014. Bdkrb2 gene −9/+9 polymorphism and swimming performance. Biol. Sport 31, 109–113.

Hittel, D.S., Berggren, J.R., Shearer, J., Boyle, K., Houmard, J.A., 2009. Increased secretion and expression of myostatin in skeletal muscle from extremely obese women. Diabetes 58, 30–38.

Keller, P., Vollaard, N.B., Gustafsson, T., Gallagher, I.J., Sundberg, C.J., Rankinen, T., Britton, S.L., Bouchard, C., Koch, L.G., Timmons, J.A., 2011. A transcriptional map of the impact of endurance exercise training on skeletal muscle phenotype. J. Appl. Physiol. 110, 46–59.

Kohn, T.A., Essen-Gustavsson, B., Myburgh, K.H., 2007. Do skeletal muscle phenotypic characteristics of Xhosa and Caucasian endurance runners differ when matched for training and racing distances? J. Appl. Physiol. 103, 932–940.

Lucia, A., Gomez-Gallego, F., Barroso, I., Rabadan, M., Bandres, F., San Juan, A.F., Chicharro, J.L., Ekelund, U., Brage, S., Earnest, C.P., Wareham, N.J., Franks, P.W., 2005. PPARGC1A genotype (Gly482Ser) predicts exceptional endurance capacity in European men. J. Appl. Physiol. 99, 344–348.

Lucia, A., Moran, M., Zihong, H., Ruiz, J.R., 2010. Elite athletes: are the genes the champions? Int. J. Sports Physiol. Perform. 5, 98–102.

Marino, F.E., Lambert, M.I., Noakes, T.D., 2004. Superior performance of African runners in warm humid but not in cool environmental conditions. J. Appl. Physiol. 96, 124–130.

Marton, O., Koltai, E., Takeda, M., Koch, L.G., Britton, S.L., Davies, K.J., Boldogh, I., Radak, Z., 2015. Mitochondrial biogenesis-associated factors underlie the magnitude of response to aerobic endurance training in rats. Pflugers Arch. - Eur. J. Physiol. 467, 779–788.

Matsuda, R., Spector, D.H., Strohman, R.C., 1983. Regenerating adult chicken skeletal muscle and satellite cell cultures express embryonic patterns of myosin and tropomyosin isoforms. Dev. Biol. 100, 478–488.

McGee, S.L., Fairlie, E., Garnham, A.P., Hargreaves, M., 2009. Exercise-induced histone modifications in human skeletal muscle. J. Physiol. 587, 5951–5958.

Mustafina, L.J., Naumov, V.A., Cieszczyk, P., Popov, D.V., Lyubaeva, E.V., Kostryukova, E.S., Fedotovskaya, O.N., Druzhevskaya, A.M., Astratenkova, I.V., Glotov, A.S., Alexeev, D.G., Mustafina, M.M., Egorova, E.S., Maciejewska-Karlowska, A., Larin, A.K., Generozov, E.V., Nurullin, R.E., Jastrzebski, Z., Kulemin, N.A., Ospanova, E.A., Pavlenko, A.V., Sawczuk, M., Akimov, E.B., Danilushkina, A.A., Zmijewski, P., Vinogradova, O.L., Govorun, V.M., Ahmetov, I.I., 2014. AGTR2 gene polymorphism is associated with muscle fibre composition, athletic status and aerobic performance. Exp. Physiol. 99, 1042–1052.

Niemi, A.K., Majamaa, K., 2005. Mitochondrial DNA and ACTN3 genotypes in Finnish elite endurance and sprint athletes. Eur. J. Hum. Genet. 13, 965–969.

Onywera, V.O., Scott, R.A., Boit, M.K., Pitsiladis, Y.P., 2006. Demographic characteristics of elite Kenyan endurance runners. J. Sports Sci. 24, 415–422.

Pandorf, C.E., Haddad, F., Wright, C., Bodell, P.W., Baldwin, K.M., 2009. Differential epigenetic modifications of histones at the myosin heavy chain genes in fast and slow skeletal muscle fibers and in response to muscle unloading. Am. J. Physiol. Cell Physiol. 297, C6–16.

Rankinen, T., Roth, S.M., Bray, M.S., Loos, R., Perusse, L., Wolfarth, B., Hagberg, J.M., Bouchard, C., 2010. Advances in exercise, fitness, and performance genomics. Med. Sci. Sports Exerc. 42, 835–846.

Rubio, J.C., Martin, M.A., Rabadan, M., Gomez-Gallego, F., San Juan, A.F., Alonso, J.M., Chicharro, J.L., Perez, M., Arenas, J., Lucia, A., 2005. Frequency of the C34T mutation of the AMPD1 gene in world-class endurance athletes: does this mutation impair performance? J. Appl. Physiol. 98, 2108–2112.

Schroder, W., Klostermann, A., Distl, O., 2011. Candidate genes for physical performance in the horse. Vet. J. 190, 39–48.

Scott, R.A., Moran, C., Wilson, R.H., Onywera, V., Boit, M.K., Goodwin, W.H., Gohlke, P., Payne, J., Montgomery, H., Pitsiladis, Y.P., 2005. No association between angiotensin converting enzyme (ACE) gene variation and endurance athlete status in Kenyans. Comp. Biochem. Physiol. A Mol. Integr. Physiol. 141, 169–175.

Sonna, L.A., Sharp, M.A., Knapik, J.J., Cullivan, M., Angel, K.C., Patton, J.F., Lilly, C.M., 2001. Angiotensin-converting enzyme genotype and physical performance during US Army basic training. J. Appl. Physiol. 91, 1355–1363.

Tringali, C., Brivio, I., Stucchi, B., Silvestri, I., Scurati, R., Michielon, G., Alberti, G., Venerando, B., 2014. Prevalence of a characteristic gene profile in high-level rhythmic gymnasts. J. Sports Sci. 32, 1409–1415.

Vaughan, D., Huber-Abel, F.A., Graber, F., Hoppeler, H., Fluck, M., 2013. The angiotensin converting enzyme insertion/deletion polymorphism alters the response of muscle energy supply lines to exercise. Eur. J. Appl. Physiol. 113, 1719–1729.

Wisloff, U., Najjar, S.M., Ellingsen, O., Haram, P.M., Swoap, S., Al-Share, Q., Fernstrom, M., Rezaei, K., Lee, S.J., Koch, L.G., Britton, S.L., 2005. Cardiovascular risk factors emerge after artificial selection for low aerobic capacity. Science 307, 418–420.

CHAPTER 12

Physiology of Training Plan: Periodization

12.1 INTRODUCTION

Periodization is a goal-oriented planning of physical training and sport performance. Current thinking on periodization is, to a great degree, based on the theoretical model of Soviet sport (Matvejev, Zaciorskij, and Verkhoshanky). The theoretical model originated from a practical need for coaches, especially in track and field, swimming, and weight lifting, to optimize the development of sport performance. Until about the 1950s these sports had just one competition period, and several factors played important roles in the structure of the annual periodization (including climate, availability of adequate training facilities, etc.). In outdoor events, the four-season climate enjoyed by several member countries of the USSR included cold and relatively long winters, and thus created a great challenge for sport preparation. Sprinting, long jump, high jump, pole vault, javelin, and discus were almost impossible to train for out of doors from November to March. Therefore, a great need existed for training periodization, including the precompetition season, the competition season, any transition period, and the off-season times.

These periods were divided into goal-setting macro-cycles, which represented a time frame that was necessary to improve performance. Macro-cycles are built from goal setting 4–10 micro-cycles. Micro-cycles are made up for each week and could involve 3–7 days of training. In many sports, top athletes train 2–3 sessions in 1 day, with each session having different goals. This pioneering work and theory synthesis of Matvejev, Zaciorskij, and Verkhoshanky were highly important for the preparation of athletes to attain optimum results. These sport scientists were instrumental in the progress of modern training theory, periodization, and sport development. As a result of their work, elite sport received international media attention and developed into a profitable business which has culminated into the most advertised and followed genre of the modern age. The more frequent events, matches, and competitions made available to the general public significantly

The Physiology of Physical Training
https://doi.org/10.1016/B978-0-12-815137-2.00012-7

 185

changed the old-fashioned structure of periodization. The approach of this chapter will be to discuss the most important physiological issues of periodization. The discussion will progress from goal setting to one training session, to the micro-cycle, then the macro-cycle, and finally training periods for a 1-year plan.

12.2 GOAL SETTING

One of the most important features of periodization and exercise training is goal setting. This is due to the fact that exercise-induced adaptation is specific and different from stress-associated adaptation. In stress-associated adaptation, different types of stimulus result in adaptive responses, which are completely independent of the type of stimulus. In other words, stress response is a nonspecific response (Selye's theory). Moreover, it is clear that exercise-related adaptation takes place in the resting period, which is very different from the stress-related adaptation. If stress and exercise-associated adaptations were the same or similar, then the training intensity and duration would be the same for short- and long-distance runners. But, this is not the case, so the brilliant stress theory of Selye is not adequately cited in exercise-associated adaptation. Exercise-induced adaptation is training and intensity dependent, which is also very different from the stress response, as discussed in Chapter 3. Intensity is the most important characteristic of training. The type of fatigue is dependent on the intensity of exercise, which initiates the adaptive response. High intensity exercise generally results in fatigue due to the accumulation of metabolites, like lactic acid, ammonia, ADP, and Pi, while exercise with low intensity and long duration is associated with depletion of glycogen stores in skeletal muscle, hyperthermia, hypoglycemia, dehydration, etc. Therefore, the intensity of exercise also influences the time which is necessary for regeneration, thus directly affecting the balance between the training and resting periods. To establish the required intensity of the exercise that will bring about the desired changes, the goal of the training must be set. To increase speed of movement, explosive strength/high intensity exercise must be used. On the other hand, if one wishes to increase endurance, the intensity must be moderated.

Just as one sets the target address on the GPS in a car, before departure, one must set the goal for training. Without a defined destination, one would not arrive at the desired spot, and without setting training goals, the effectiveness of the training would be negligible. Therefore, goal setting is one

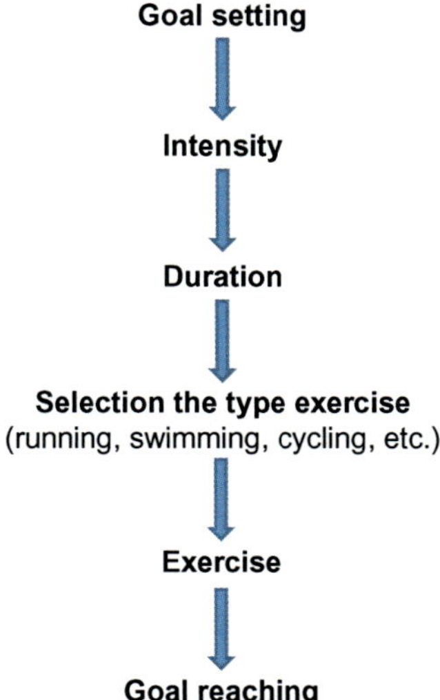

Fig. 12.1 Theoretical model of goal reaching in sport. Goal setting before exercise session is important, because it directly affects the intensity of exercise, which then determines the duration of the loading. Then we can select the type of exercise; for instance, to increase aerobic capacity, we can select running, swimming, or cycling. The repetition of this exercise after a period of time cause adaptive response and the goal of exercise (or periods of exercise) can be achieved.

of the most important and difficult tasks for the coach. Knowing the aim, the coach must select the proper intensity and type of exercise, while the necessary rest intervals would be dependent on the intensity (Fig. 12.1). To set an optimal, individual, sport-specific goal is not an easy task. For example, elite athletes (most probably to reach maximal performance), juniors (to provide the best age–associated development), recreational athletes (health promotion), and athletes rehabilitating from injury (increased regeneration and injury prevention), would all have different goals, and therefore different training protocols.

In order to set optimal individual goals, the coach or trainer must collect as much information as possible on the athlete. Firstly, one must know the physiological requirements of the sport. Without this knowledge it would not be possible to design training plans or establish goals. It would be foolish to think that one only needs aerobic endurance to excel at soccer since the duration of the game is 90 min. The majority time of a soccer match is covered by low intensity running at about 70% of VO_2max and top players have 150–250 brief intense actions, showing a need for anaerobic alactic and lactic training as well. In addition, the position one plays is also crucial.

Usually central midfielders control the ball for the longest period in the game and generally they cover the longest running distance as well (Bangsbo et al., 2006). A full-length judo match can result in blood lactate levels higher than 12 mmol/L with an increase in the levels of extracellular markers of muscle adenine nucleotide catabolism, urea, and creatinine, suggesting a high level of metabolic demand (Degoutte et al., 2003). During fencing the blood lactate levels can be around 7 mmol/L (Milia et al., 2014), suggesting the need for aerobic and anaerobic exercise training in order to perform at a high level. Being aware of the requirements of a sport is therefore crucial for the establishment of appropriate goals for training sessions.

In addition, it is important to create sport-specific tests that correlate well with the performance of a given sport. The results of these tests permit one to elucidate limiting factors of individual performances. For elite athletes to reach peak performance, it is crucial to find the weakest physiological or technical characteristic and set it as a primary target for development and improvement. Without question, to be able to set the optimal goals of the athletes, objective, measured information related to physical fitness, technique, and even psychological conditions are necessary. In this book, the author focuses on the physiological limits, while at the same time recognizing that sport performance is a result of complex physiological, mental, and cognitive interactive processes.

12.2.1 Structure and Physiology of One Training Session

One training session might be the simplest unit of exercise training periodization. As noted earlier, the key point is to define the goal of the training process. In most sports, concurrent training (which means the various types of training used to reach the targeted performance) exists. One training session can be built with different exercises, with different goals; therefore, it is not easy to set the goal for the training session. Is it possible to have exercise bouts with different goals, e.g., to improve aerobic capacity, anaerobic lactic capacity, and muscle hypertrophy, in the same training session? The answer is probably yes. There are many transfer effects between different types of exercise, but all are the result of muscle contraction. The interference between different types of contraction could lead to different sum effects from each individual exercise. In one training session the order of these different types of exercise could also produce different results. As an example, 10×30 m running with maximal speed at the beginning or at the end of a 3-h training session could have different effects. If it is done at the beginning it can lead to the development of maximal speed, while when used at the end of a training session, it can improve speed endurance. Moreover, high intensity exercise requires anabolic hormones to bring about increased protein synthesis, but the secretion of testosterone is dependent upon the order of the exercise (Di Blasio et al., 2016; Vuorimaa et al., 2008). It has been shown that if strength training is followed by endurance exercise, the level of circulating testosterone is significantly reduced 24 h after training, whereas when the subjects do endurance training followed by strength training (Schumann et al., 2013), the results are dissimilar. Generally, high intensity resistant exercise might be less effective after long-term aerobic exercise, which can decrease the levels of testosterone (Fournier et al., 1997).

The definition of the goal of the exercise helps to establish the intensity of the exercise. Intensity is the key factor which influences the adaptive response. For example, if the goal of the exercise is to cause muscle hypertrophy, 80% of the maximal load, with seven repetitions and seven sets would be appropriate loading. If the goal of the exercise is to improve acceleration ability in soccer players, one could select running from a standing position with maximal sprints using a sled pull (Fig. 12.2A), or use vertical jumps at 40% of maximum load with maximal speed for five repetition in six sets (Fig. 12.2B). As an alternative, one can use repeated throws of 10–15 kg weights (kettle bell) with maximal effort (Fig. 12.2C), or running upstairs at maximal speed (15×20 m) (Fig. 12.2D). These exercises increase the explosive strength of legs, which is crucial for the acceleration of soccer players.

So, goal setting helps to establish the intensity of the exercise and the intensity of the exercise determines the duration of the exercise, demonstrating the inverse relationship between intensity and duration. The intensity of the exercise and the intensity of the whole training session would also influence the adaptive response. The main physiological responses of different types of exercise are summarized below.

Acute Exercise at Maximal and Submaximal (>85% of Maximum) Intensity Will Bring About the Following Physiological Responses

Generally, high intensity interval exercise, like exercise with Tabata protocol which consists of eight sets of exercise at the intensity of 170% of VO_2max with 10-s rest (Tabata et al., 1996), results in enhanced mRNA expression, especially of those proteins that are involved in mitochondrial biogenesis and/or enhanced protein synthesis (AMPK, PGC-1a, SIRT1, NRF1, NRF2, TFAM, Akt, mTOR, HIF1a (Abe et al., 2015; Terada and Tabata, 2004) in skeletal muscle. Dr. Izumi Tabata, currently a professor

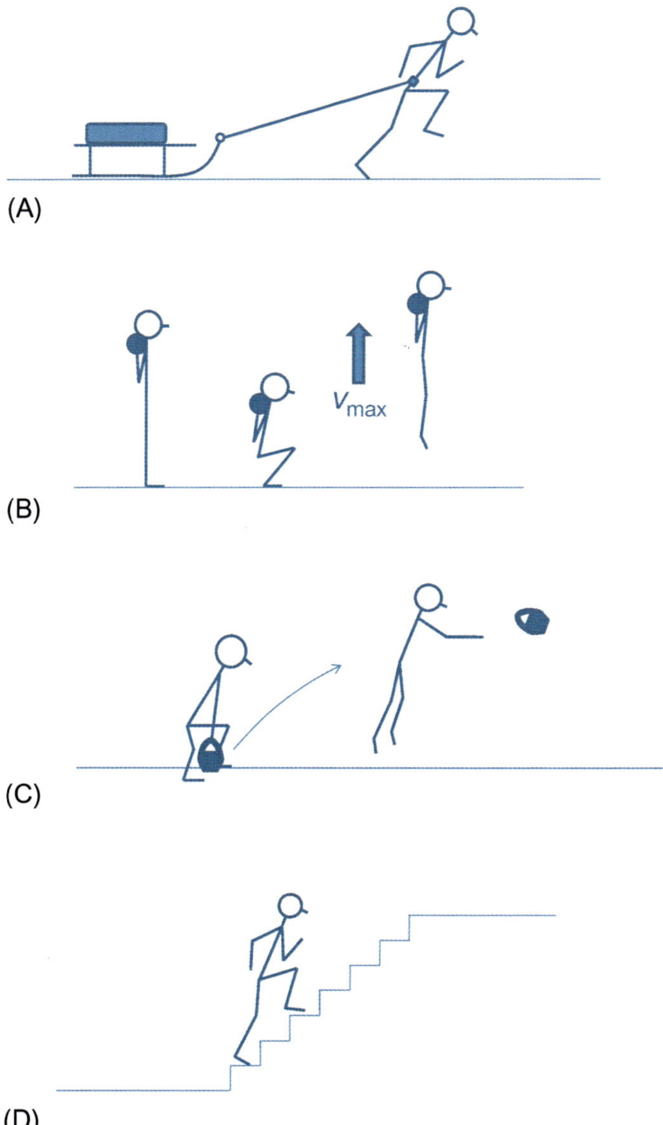

Fig. 12.2 Model to reach the same goal with different type of exercise. The explosive strength of the quadriceps femoris can be improved by different types of exercise, such as sled pull (A), vertical jumps (B), kettle bell throw (C), or running upstairs (D). It is advised to use different types of exercise to enhance complex adaptive response.

at Ritsumaker University, Japan, was one of the first scientists to use and describe the effects of high intensity interval training.

Exercise with intensities over 130% of VO_2max increases oxygen flow to muscle mitochondria by about 100-fold, and despite a large increase in blood delivery to contracting skeletal muscle, partial hypoxia takes place. This increase in metabolic demand cannot be covered by adequate energy supply, leading to increased levels of AMP, lactic acid, Pi, and other metabolites. For improved oxygen delivery, HIFa and VEGF induction would initiate the development of enhanced vascularization, especially with active recovery between exercise bouts (Wahl et al., 2014). It is important to note that active rest between exercise sets would further enhance the effects of this exercise regimen compared to passive resting periods, by enhancing lactate clearance and glycogen resynthesis (Choi et al., 1994). Similar findings have been reported when the effects of four sprints at maximal effort of 300, 3×100, 400, and 2×200 m were done by young athletes. The applied methods all resulted in increased elevation of lactic acid, but intermittent workouts (3×100, 2×200 m running) appeared to be superior when compared with continuous workouts of the same total distance by increasing the ability for energy production via the lactate system (Saraslanidis et al., 2009).

When the acute results of high intensity interval cycling and high-volume cycling of well-trained cyclists were studied for their hormonal profile, it turned out that high intensity work increased circulating growth hormone and cortisol levels compared to the high-volume exercise. High intensity-associated acidosis could be important in the elevation of these hormones, as the hormones elevate protein synthesis, protein turnover, repair, regulation of metabolic function, etc. Acute high intensity exercise results in elevated levels of circulating microRNA, which could be important as regulators of vascular and anabolic adaptive responses (Kilian et al., 2016).

High intensity resistance exercise results in elevated circulation of testosterone, suggesting an enhanced anabolic process. However, it must be mentioned that longer resting periods between sets, such as 3 min, is superior to short 1-min rest periods for the body to enjoy the long-term effects of elevated testosterone (Scudese et al., 2016). In addition, active rest with high acute intensity eccentric exercise readily cause micro damage if sarcomeres leading to muscle soreness (see Chapter 3 for a detailed description of muscle soreness).

Exercises that enhance cardiovascular function can beneficially affect motor learning, which could be important in the enhancement of sport performance. The results of a recent study indeed reveal that high intensity interval exercise improves the learning of ballistic tasks in trained and untrained limbs by reducing interference between trained and untrained hemispheres in the brain (Lauber et al., 2017). The authors also suggest that high intensity interval exercise could contributes to cognitive health, as it alters the event-related potential measured by electroencephalography (Kao et al., 2017). These results suggest that certain effects of exercise on brain function are dependent on the intensity of the exercise.

Interestingly, acute high intensity exercise could have a beneficial impact on patients with type 1 or type 2 diabetes. This kind of exercise results in increased production and accumulation of lactic acid, which is considered to preserve cognitive function and awareness of hypoglycemia (Rooijackers et al., 2017); moreover, the production of nitric oxide metabolites with this intensity of exercise could be important to the health-promoting effects of Tabata protocol.

If We Use Single Bouts of Exercise With 60%–80% Intensity, We Can Expect the Following Physiological Responses

Single bouts of endurance exercise with an intensity of 60%–80% of maximum are often used to target enhancement of cardiovascular function, including stroke volume (Wang et al., 2012). Since carbohydrate is the main energy source of exercise with this intensity, a significant decrease, about 30%, in muscle glycogen content can be observed after 4 h of exercise (Gejl et al., 2014). Marathon running can decrease the muscle glycogen content by more than 40% of the prerace value and induce a wide range of cellular signaling cascading that is necessary to cope with this challenge and does so via mitogen-activated protein kinase (MAPK) pathways (Yu et al., 2001). Moreover, it is clear that marathon running results in changes in circulating miR-levels, and these miRs are important regulators of the adaptive response of skeletal muscle and other organs (Baggish et al., 2014).

When the maximal force of knee extensors, sarcoplasmic function, and muscle glycogen levels were studied in young elite soccer players immediately after and then 24 h following a soccer game, the data revealed that maximal force decreased 11% and 10%, respectively (Krustrup et al., 2011). The calcium re-uptake of sarcoplasmic reticulum was also suppressed immediately after the game, but recovered after a day of rest, and the muscle glycogen was 57% and 27% lower right after and 24 h after the game compared to pregame values (Krustrup et al., 2011).

Resistance exercise with an intensity regime is appropriate to cause muscle hypertrophy and there is ample data using single bouts of resistance exercise at 60%–80% of maximal load which show increased levels of circulating total and free testosterone (Kraemer et al., 1998). Resistance training at 60%–80% of maximum for 1 RM often results in enhanced secretion of growth hormone in the rest period following the exercise session, which suggests that an elevated anabolic process has taken place. Indeed acute leg extension with 75% of max with short 1-min and long 5-min rest periods between sets resulted in different levels of testosterone and synthesis of myofibrillar proteins (McKendry et al., 2016), underlining the important role of the loading-rest period balance to bring about adaptive responses. Changing either of these factors could have an impact on adaptation.

The effects of exercise are systemic (Radak et al., 2008), which means that most of the organs are affected by the exercise-induced changes of metabolic, hormonal, and neuronal systems. As an example, it has been shown that the cerebral oxygenation of middle-distance runners from Kenya, running 5% faster than their normal race speed for a 5 km race, were found to decline, which correlated well with the onset of fatigue (Santos-Concejero et al., 2017). Overall, it main difference between the effects of exercise bouts with 60%–80% of the maximal intensity is due to the differences on the accumulation of metabolites, depletion of energy sources, recruitments of fiber type, and hormonal secretion. These differences would result in different regeneration after exhaustion after exercise bouts with higher and lower intensities than 80% of maximal intensity.

If One Uses a Single Bout of Exercise With Moderate (<60% of Maximum) Intensity, the Following Physiological Response Can be Expected

Endurance exercise bouts with less than 60% VO_2max are termed moderate aerobic exercise, and when carried out for a long duration, these have been shown to cause adaptive responses. One of the targeted adaptive responses could be an improvement in thermoregulation. Endurance exercise with 50%–60% of VO_2max with a duration longer than 1 h easily increases core temperature and results in a significant loss of fluid through sweating (O'Neal et al., 2012). Prolonged exercise of this intensity, like marathon and ultra-marathon running, triathlon, etc., results in a significant depletion of glycogen stores in skeletal muscle. Exercise of this intensity and duration utilizes fat as the primary energy source and contributes to a more efficient use of free fatty acids and a reduced rate of glycogen use. Indeed, reduced carbohydrate oxidation is associated with decreased pyruvate production and muscle glycogen content (Mourtzakis et al., 2006). During recovery from exercise bouts of this moderate intensity, protein degradation takes place, as is evidenced by a net release of threonine, tyrosine, methionine, and lysine from skeletal muscle, a phenomenon which is likely important in re-establishing homeostasis in extra-muscular tissues (Mourtzakis et al., 2006). Intramuscular triglycerides could serve 20%–25% of the energy during moderate intensity exercise, when the duration is longer than 1 h (van Loon et al., 2001), although these droplets just contain a very small percentage of whole-body lipid stores. Indeed, it has been shown that exercise at this moderate intensity causes a shift from carbohydrate metabolism to fat as the main energy source, especially during the last part of the prolonged exercise. Ketone bodies are alternative fuel sources produced by liver from fatty acids, in conditions where the carbohydrate availability is decreased, like prolonged physical exercise, free fatty acids are the primary substrate for ketogenesis and circulated ketones can be uptaken by skeletal muscle, and the ketonic transfer to type I fibers are greater than type IIa fibers and correlates well with the oxidative capacity of the fibers. The utilization of ketone bodies is dependent on the level of physical fitness and trained individuals can use and uptake greater amount of ketones. Moderate intensity exercise with a long duration results in postexercise increase in ketone bodies (Evans et al., 2017). During exercise of this intensity, such as in ultra-marathon running, the body has to cope with a sustained challenge. It has been suggested that the performance of ultra-marathon running is closely linked to the level of VO_2max and the mitochondrial oxidative capacity of skeletal muscle (Millet et al., 2011). Acute bouts of exercise of this intensity and long duration often lead to oxidative damage of macromolecules (Radak et al., 2003). However, it is important to note that regular exercise with this moderate intensity results in enhanced activity of antioxidant systems and decreased levels of oxidative damage.

As stated earlier, the effects of exercise are systemic, and another example shows that exercise of this moderate intensity beneficially effects brain function (Guillou, 1989). Some studies on animals showed that voluntary running has more benefits on spatial memory of rats than forced exercise.

Resistant exercise with a load lower than 60% of 1 RM is generally used to increase strength endurance, while it is important to note that intensity in resistance training is dependent not just on the size of the load but also on the speed by which one moves. Therefore, 60% of 1 RM with more than 20 repetitions would result in improved strength endurance with enhanced utilization of slow-twitch fibers, while four to six repetitions with the same load with maximal executional speed would activate fast-twitch fibers and improve explosive strength. Therefore, strength endurance exercise could cause increased mitochondrial density in mixed fibers and the recruited fast-twitch fibers. It is important to note that increased intensity does not induce mitochondrial biogenesis in slow-twitch fibers, but does in fast-twitch fibers, while increasing the duration of the exercise promotes enhancement only in slow-twitch fibers (Bishop et al., 2014). Strength endurance training would further induce vascularization of working muscle by the activation of angiogenic growth factor (Gavin et al., 2007), resulting in a more efficient capability to utilize energy and oxygen. The increased number of satellite cells after acute exercise at this load, with high repetition, could be important to regeneration and remodeling.

On the other hand, explosive exercise could activate adaptive responses in fast-twitch fibers, leading to a wider range of neuromuscular adaptation (Linnamo et al., 2000). It is interesting to note that temperature plays a role in power performance and secretion of growth hormone in both genders. Comparing performance and growth hormone levels after exercise bouts in temperatures of 20 and 30°C, it seems that higher temperatures facilitate power and growth hormone levels (Casadio et al., 2017). However, it is important to note that resistance exercise with a load of 60% or less would not trigger significant increases in the production of anabolic hormones.

12.2.2 The Structure and Physiology of a Micro-Cycle

The micro-cycle is a convenient unit of periodization as it generally covers a 1-week period. This period is not enough to cause meaningful and measurable adaptive responses, but it is widely used as a convenience. Moreover, it is impossible to summarize the physiological effects of a micro-cycle due to the differences in training goals for each one. Every micro-cycle has a goal, which demonstrates the outcome of the goals of each training day in the given micro-cycle. In order to have regularity of training, each week must have at least 3 days with training. For those who exercise with the aim to improve health, three training sessions per week could be effective, but with an increased level of physical fitness and with highly competitive goals, one could have as many as 10–15 training sessions in a micro-cycle. The sum load of a weekly micro-cycle is dependent on the goals of daily training sessions. A micro-cycle that aims to increase aerobic performance must have a longer sum duration than the micro-cycle that aims to improve speed. The sum load of each micro-cycle is given by loads of each training sessions in the micro-cycle. In general, not every day in the micro-cycle has the same load, since not every day in a micro-cycle has the same goal. The training load for each day is dependent upon the macro-cycle goal. The loading structure of the micro-cycle is also influenced by the macro-cycle and the period in which the micro-cycle is located. Moreover, the loading structure of the micro-cycle must be personalized to fit to each athlete. The average load of the training session is expressed as volume x intensity. The intensity of the exercise is expressed as a percentage of a given type of exercise such as percentage of 1RM in resistance training and percentage of VO_2max or heart rate in endurance training. The intensity of endurance exercise is also judged by the accumulation of lactic acid, having three thresholds. Based on the lactate shuttle theory developed by George Brooks (1986), lactate is also produced during fully aerobic conditions, rest, or moderate intensity exercise. During moderate intensity exercise the lactate turn point (LTP1) is the level of lactate from where lactate concentration increases in the circulation with the matched increase in ventilation. Exercise at this intensity can be done for a very long duration such as an ultra-marathon, iron man triathlon, or ultra-distance swimming. This is accomplished because of the steady-state level of lactate levels due to the balance of lactate production and elimination. However, when someone reaches the second threshold (LTP2), the lactate levels would abruptly increase with the matched elevation in ventilatory parameters (see Chapter 5).

Although there are a number of existing problems to assume the proper intensity of endurance exercise, setting the most adequate duration of the exercise is even more problematic (Hofmann and Tschakert, 2017). Duration can be expressed as time, distance, set, repetition number, etc. It is known that as one increases the distance in cyclic sports (running, swimming, kayak, etc.) the average speed declines. The average speeds of the 100 and 200 m dash world records of Usain Bolt were almost the same, at 37.58 and 37.51 km/h, respectively (this is due to the fact that the acceleration of the 100 m starts from the upright position, while in the 200 m the runners arrive at high speed for the second 100 m), whereas as the races get longer, the speed decreases (43.03 s for the 400 m sprint and 33.3 km/h for the 10,000 m where the average speed is 22.8 km/h). This correlation between speed and distance (the maximal distance can be set to every speed and one can express the given distance as a percentage of maximal distance of given speed) is relevant. The time of 2:02.57 is, at the time of writing, the best time in the marathon and this corresponds to an average speed of 20.0 km/h, so for 20 km/h speed 42.19 km is the maximal distance; if the runners run 21 km with 20 km/h speed, this will correspond to 50% of the maximal distance.

In general, older athletes have slower rates of recovery after training sessions than younger athletes, and this must be reflected in the loading structure of the micro-cycle (Fig. 12.3A/B). Therefore, when one creates the goals for each training day in the micro-cycle, one must pay attention to the fact that training and rest together result in adaptation. In other

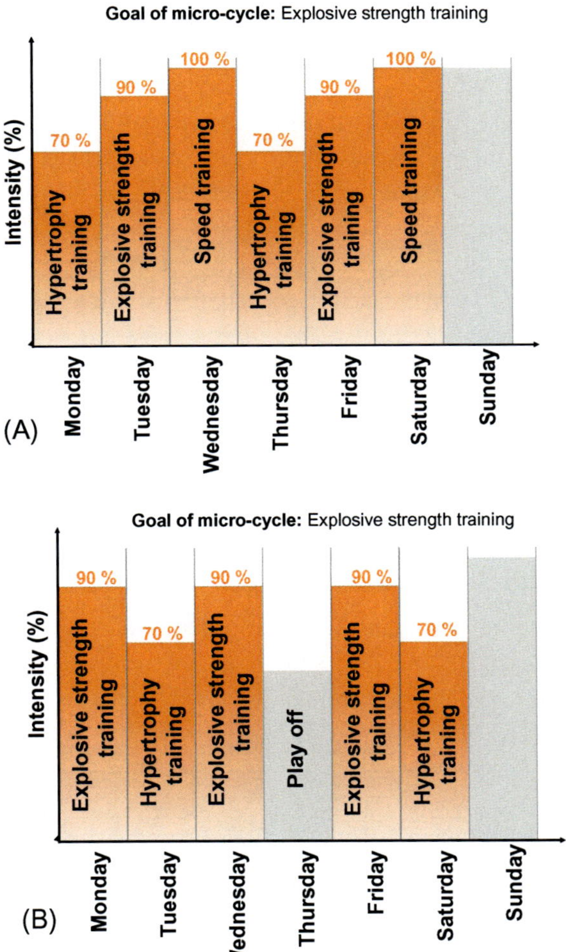

Fig. 12.3 The loading structure of a micro-cycle is dependent on the regeneration capacity of the athletes. There are many factors which influence the loading structure of micro-cycle, and the age of the athlete could be one. Younger athletes generally have better regeneration capacity than older athletes, therefore they can endure greater loads with shorter periods of rest. Panels show microcycles with the goal of strength training. Panel A is for an athlete with great and panel B is for an athlete with moderate recovery capacity.

words, most of the athletes cannot endure maximal loading each day in the micro-cycle. Therefore, it is wise to set up the goals in this cycle to allow necessary regeneration time for the athletes to achieve the targeted goal of the micro-cycle. One micro-cycle usually consists of exercise trainings with different goals, and a 1-week training period usually does not result in measurable adaptive responses on trained individuals.

12.2.3 The Structure, Function, and Physiology of the Macro-Cycle

The macro-cycle is one of the most important units of periodization. The reason is because measurable adaptive responses take place after 6–8 weeks of exercise training. In some situations, 4 or 12 weeks of micro-cycle could be appropriate. The main task of the macro-cycle is to achieve the target goal, which could be an increased level of aerobic or anaerobic capacity, improved level of maximal or explosive strength, or an enhanced level of acceleration capacity, among others. The macro-cycle starts with the measurement of the level of the targeted goal. An example is acceleration capacity, which can be measured by running time. However, because acceleration is strongly dependent on explosive and maximal strength, it is highly advised to measure the performance of the long jump from a standing position, kettlebell throwing forward, and/or snatching. In order to test the efficiency of the macro-cycle, the same test must be applied at the end of the cycle. Therefore, the main goal of the macro-cycle is to cause an adaptive response of the targeted condition, ability, or coordination. The pre- and postmeasurements of the macro-cycle show whether we have reached the goal.

Macro-cycles are made up from micro-cycles. The sum goals of the accumulated micro-cycles should result in improved performance in the characteristic which was the goal of the macro-cycle. Depending on the goal and location (preoperational period or competition period) of the macro-cycle, one, two, or three micro-cycles with the load over 80% can be followed by a micro-cycle with a moderate load of 60% or less. The pattern of personal regeneration should be considered to set up the loading structure of the macro-cycle.

Recently, block periodization has been used to provide a challenging structure for better enhancement of performance. During block periodization, the loading is concentrated towards improving qualities, but with similar intensities. For instance, if the goal of the macro-cycle is to improve anaerobic performance in a 12-week program, one can have two high intensity interval training periods each week, or one can use block periodization which means that in the first 3 weeks, five sessions of high intensity interval training followed by only one weekly session can be used. Block periodization was found to be superior to the same weekly loading program by the relative improvement in VO$_2$max, power output at 2 mmol/L lactate level (Ronnestad et al., 2014). Fig. 12.4 shows an example of block periodization, used efficiently for endurance and resistant periodization (Bartolomei et al., 2014; Ronnestad et al., 2016). Besides block periodization, training load

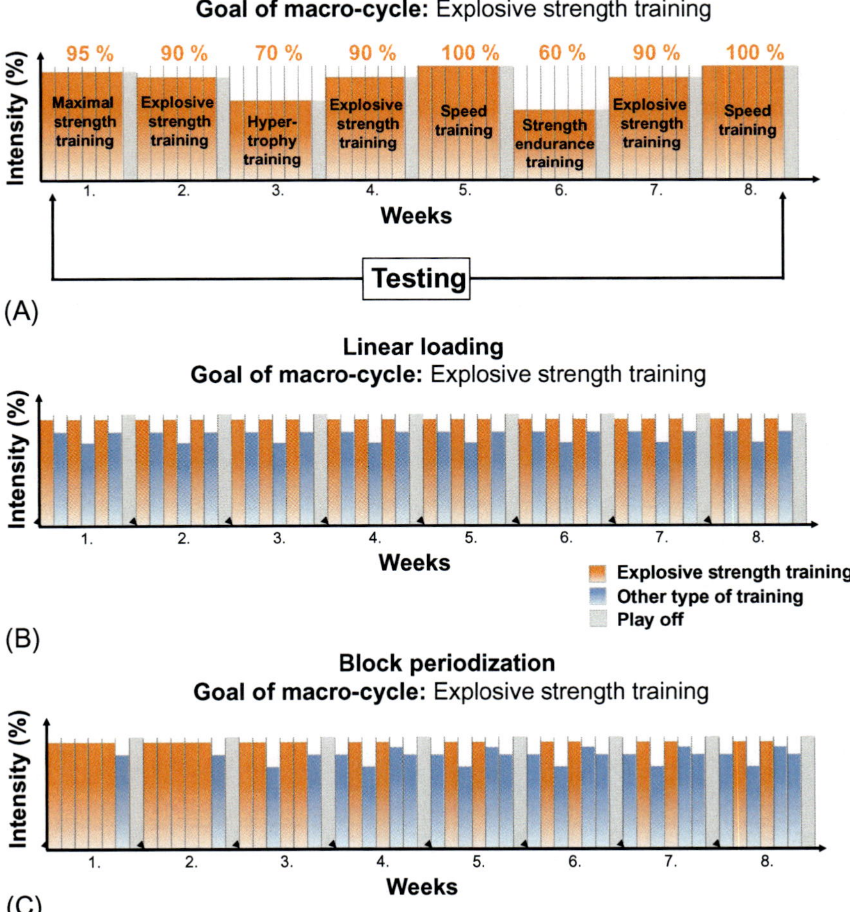

Fig. 12.4 Goal setting and loading models of the macro-cycle. The goal of the macro-cycle is reached by the sum effects of goal-oriented micro-cycles (A). The distribution of loads can affect the effectiveness of adaptive response. Block periodization (C), when the exercise load of targeted ability (explosive strength in this figure) is given at a larger degree in the first 2–4 micro-cycles, can be superior in the same abilities to linear periodization (B), where the load is distributed equally in each micro-cycle.

can be given by so-called nonlinear periodization, including an undulating model, in which the volume and intensity are changed more frequently, perhaps on a daily or weekly basis.

12.2.4 Macro-Cycles for Aerobic Endurance

One of the main goals of macro-cycles was to improve aerobic capacity to target the improvement in cardiovascular function, including greater stroke volume, arterial-venous oxygen difference, mitochondrial density and capillarization, along with increased oxidative metabolism of fat and carbohydrates. Because of the specificity of exercise-induced adaptation, different intensities and duration are optimal to increase stroke volume or capillarization of skeletal muscle. It is not easy to assess all of the components that influence aerobic capacity, but it is strongly advised to perform laboratory and/or field tests to assess VO_2max, ventilatory markers, and possible measurements of metabolism before and after each macro-cycle (see Chapter 13).

Intensity is the key regulator of exercise-induced adaptation. For aerobic endurance, intensity is often given as the percentage of VO_2max. But, because of the convenience of the very strong relationship between heart rate (HR) and VO_2max, HR zones are often used to define intensity during training. If heart rate is used, generally three intensity zones are distinguished: (1) light intensity below ventilatory threshold (V1); (2) moderate intensity when the exercise is done between V1 and above the compensatory ventilator threshold (V2); and (3) high intensity above compensatory ventilator (V3). When the effects of 6 months of training were tested (called a macro-cycle, but it included periodization and competitive periods as well) on the Spanish national middle and long-distance endurance runners, it turned out that 71% of total training volume was spent in the low intensity zone, V1 (below 60% of VO_2max), 21% in the moderate V2 zone (60%–85% of VO_2max), and 8% in V3, the high intensity zone (over 85% of VO_2max) (Esteve-Lanao et al., 2005).

Eight weeks of supra-lactate threshold training by long-distance runners, with the goal to improve aerobic energy cost of running and stride characteristics, have been studied. Stride rate variability and running energy cost decreased, while VO_2max increased (Slawinski et al., 2001), suggesting that the adaptive response included cardiovascular and skeletal muscle-related changes. University-level cross-country runners performed 8 weeks of training with 97 km weekly total running distance, which included 9% interval training after the summer holidays. The running performance at

8 km improved by 2.8% and the results of muscle biopsy samples revealed that the diameter of slow-twitch fibers decreased by 3%, while the peak force of slow-twitch fibers increased by 18% and mixed fibers (MHC IIa) by 11% (Harber et al., 2004). Moreover, the single slow-twitch fiber contraction velocity and power decreased by 23% and 20%, respectively, indicating that the gain in performance was probably due to increased cardiovascular effects of training rather than muscular adaptation. Competitive rowers performed a 3-month training program, which was nearly exclusively dedicated to endurance activity (endurance, 12.6 ± 0.7 h/week and strength, 1.0 ± 0.9 h/week). After this period the left ventricular mass and resting stroke volume increased significantly, resulting in enhanced cardiovascular function (Weiner et al., 2015).

When an 8-week intervention, with two sessions of resistance training per week, was included into the macro-cycle of well-trained middle-distance runners, it turned out that the resistance training could prevent a decrease in stride length (cm)/speed (m.s) (SLS) ratio, which normally happens during fatiguing running bouts (Esteve-Lanao et al., 2008). In addition, when explosive and plyometric training was added to the 12-week micro-cycle of ultra-marathon runners, the running cost decreased significantly (Giovanelli et al., 2017). The economy of exercise is crucial to top performance. Unlike VO$_2$max, where improvement is limited, there are data to show that professional athletes could improve the economy of exercise throughout their career. The former women's marathon world record holder Paula Radcliffe (2:17:42) improved her running economy between 1992 and 2003, and used significantly less oxygen to maintain a running speed of 16 km/h (Lundby and Robach, 2015). The economy of running is defined as the energy demand for a given velocity of submaximal running, and is determined by measuring the steady-state consumption of oxygen and the respiratory exchange ratio.

Strength training programs result in better utilization of elastic energy, more efficient recruitment of motor units, and a reduction in the amount of energy wasted in braking forces. The addition of a heavy strength training program to the normal endurance program also improved the economy of cycling and fractional utilization of VO$_2$max and caused fiber type shifts from type IIAX-IIX to type IIA (Vikmoen et al., 2016). Unfortunately, the molecular background of improved efficiency is not very well known. However, higher levels of slow-twitch fibers could be an advantage since they are more economical than fast-twitch fibers and possess a greater number of mitochondria, which could also promote greater economy.

University-level rugby players participated in an 8-week training program with five strength training sessions and four field-based training sessions per week. Before and after training, the macro-cycle test was used to measure VO_2max and ventilatory markers. Results revealed that the improvement of peak VO_2 during 8 weeks of rugby training may have been caused by muscle O_2 supply, rather than increased CO or muscle O_2 extraction (Takagi et al., 2017). Similarly, good effects of plyometric training versus aerobic training were found for university-level rowers. Four weeks of plyometric training improved 500 m rowing performance and moderately increased peak power (Egan–Shuttler et al., 2017).

McConell et al. (1993) studied whether a 4-week reduction in running distance (24%) and intensity (30%) of the accustomed training distance of well-trained distance runners would significantly alter the running performance (McConell et al., 1993). The results indicated that although the VO_2max was not significantly altered, the running performance decreased in most runners, suggesting that training intensity during the reduced training periods was important for the maintenance of 5 km running performance (McConell et al., 1993).

In summary, it appears that a well-designed macro-cycle with the goal of enhancing aerobic endurance capacity can cause increased stroke volume, VO_2max, oxygen utilization, and improved economy of the exercise.

12.2.5 Macro-Cycles for Anaerobic Endurance, HIIT, and Tabata Protocol

Six weeks of HIIT training on well-trained endurance runners has been shown to result in increases in maximal treadmill speed, whereas plasma lactate decreased after HIIT. The cross-sectional area of type II fibers tended to decrease, while no changes were observed in VO_2max, fiber type, capillary density, citrate synthase, and 3-hydroxyacetyl CoA dehydrogenase activities. Lactate dehydrogenase (LDH) activity increased in homogenates ($P < .05$) and type IIa fiber pools (9.3%, $P < .05$). The change in the latter correlated with an absolute interval training speed ($r = .65$; $P < .05$). In conclusion, HIIT in trained endurance runners caused no adaptations in muscle oxidative capacity, but increased LDH activity, especially in type IIa fibers is seen in relation to absolute HIIT speed (Kohn et al., 2011). When cycling exercise sessions consisted of $8–12 \times 60\,s$ intervals at approximately 100% of peak power, 2 weeks of training resulted in massive increases in mitochondrial biogenesis-involved proteins, and upregulation of CS and cytochrome c oxidase (COX) activities in skeletal muscle of healthy men were observed

(Little et al., 2010). HIIT uses carbohydrate as a main energy source, and causes increases in GLUT4 content in order to feed the working muscle with glucose. However, HIIT alters fat metabolism as well. It has been shown that 6 weeks of cycling at 90% of VO_2max, with 10 4-min exercise bouts, significantly increased the body fat metabolism and the levels of fat transport proteins in sarcolemma and mitochondria (Talanian et al., 2010). Indeed, HIIT significantly increases postexercise energy expenditure, which can lead to loss of body fat. However, it must be noted that exercise with moderate intensity (below 50% of VO_2max) can utilize energy sources via increased circulation but at higher intensities, muscle-stored energy sources would dominate the regeneration of ATP. Nonetheless, it seems that block periodization for competitive cyclist training provides superior adaptations to traditional methods, despite similar training volume and intensity.

HIIT-associated fatigue is substrate dependent, while long-term continuous exercise-associated fatigue could be due to decreased availability of energy sources. Circulating substances can reach brain and exert central fatigue and the training-associated adaptation increases the central fatigue resistance, which contributes to enhanced high intensity exercise endurance capacity, allowing greater performance to be extruded from the muscle (O'Leary et al., 2017). As an example, 16 weeks of exercise training resulted in more efficient cerebral oxygenation during cortical activation compared to a nonexercised control group. Furthermore, HIIT and moderate continuous training were superior to resistance training for task-efficient cerebral oxygenation and improved oxygen utilization during cortical activation in older individuals (Coetsee and Terblanche, 2017). In addition, HIIT training, rather than continuous training, has a strong potential to improve brain-derived neurotrophic factor and glial cell line-derived neurotrophic factor through a greater increase in H_2O_2 and TNF-α as oxidative stress and pro-inflammatory factors (Afzalpour et al., 2015).

12.2.6 Macro-Cycles for Muscle Hypertrophy

To induce large increases in muscle mass, the cellular mTORC1-Akt pathway activation could be essential. The muscle tension-induced signaling is activated with micro damage, and this signaling pathway is then supported by an elevated secretion of anabolic hormones. Key initial cellular pathways for hypertrophy are localized in the skeletal muscle and are supported by circulating anabolic hormones. Activation and proliferation of satellite cells, by nitric oxide, could also be necessary to increase the number of nuclei in skeletal muscle (Leiter et al., 2011). Satellite cells are the source of nuclei from regenerating and developing skeletal muscle. The orchestral response of the system is controlled by microRNA-s and these posttranscriptional regulators are involved in muscle hypertrophy (Koltai et al., 2017).

Undulating or nonlinear periodization is often used to enhance muscle size. With this methodology, the intensity, duration, exercise choice, and type are selected with relative freedom. The exercise-induced elevation of circulating testosterone is associated with the size of the contracting muscle groups, and significant differences have not been discovered in either high volume (3×10–12 reps, 1 min rest) or high intensity groups (13×3–5 reps, 3 min rest) (Mangine et al., 2017). When a 12-week resistance training program was applied at greater than 60% of 1 RM, training resulted in fiber hypertrophy with proportional increases in fiber P_o and absolute power (Widrick et al., 2002). It is important to note that when the external work load is normalized by the number of repetitions at 70% of 1 RM or 35% of 1 RM, the total muscle glycogenolysis is the same, regardless of the intensity of the resistance exercise (Robergs et al., 1991). The need for an enhanced level of protein synthesis during hypertrophy might require an increased network of ribosomes, where the amino acid assembly takes place on the code carried by mRNA-s. It has been shown that 8 weeks of resistant training (70%–90% of 1RM) using leg press, knee extension, and knee flexion exercises resulted in activation of pro-synthetic cellular signaling pathways (mTOR, mitogen-activated protein kinase/extracellular signal-regulated kinase (MAPK/ERK), increased cross-sectional area of the muscles, and increased ribosomal biogenesis (Figueiredo et al., 2015). MAPK/ERK phosphorylates activate transcription factors that translocate to the nucleus and bind to special sequences of DNA to promote or block the recruitment of RNA polymerase. This phenomenon either stimulates or blocks assembly of mRNA, which codes the sequence of proteins that are important to adaptation. MAPK/ERK activation and subsequent muscle hypertrophy, along with an increased number of repetitions at 70% of 1 RM, have been observed after 20 weeks of variable resistance training (Walker et al., 2013). Both the variable resistance group and the constant resistance groups performed the exercise on cam-based devices. The former group achieved better performance and greater activation of testosterone (Walker et al., 2013).

12.2.7 Macro-Cycles for Explosive Training

This type of training is often called strength-power periodization because the main goal is to increase the power of the exercise by increasing the speed. Many agree that stronger athletes tend to develop maximal force faster.

However, the development of explosive strength must be sport-specific. Hammer throwers use heavier loads than javelin throwers, because during the rotation of the throwers, the weight of the hammer results in force greater than 3000 newton on the thrower. Therefore the maximal force of the hammer throwers is significant and they have to use heavier weights than javelin throwers to develop their explosive strength.

When soccer players were subjected to strength-power periodization (involving half squats or jump squats, depending on the respective training block) or so-called optimum power load (jump squats at the optimum power load) for 6 weeks, three times a week, the optimal training load was more effective in improving sprint and jump performance (Loturco et al., 2016).

Interestingly, when the effects of 8 weeks' training of free weights and training using pneumatic equipment were compared in healthy men, the pneumatic training group showed slightly greater increases in maximal velocity and power (Frost et al., 2016). It is possible that training with pneumatic resistance may offer advantages if attempting to improve power at lighter, relative loads by affording an opportunity to achieve consistently higher accelerations and velocities (Frost et al., 2016). However, it must be mentioned that kettlebell swing and especially throw are very effective exercises to enhance explosive strength (Lake and Lauder, 2012). Indeed, kettlebell swings with a load of 16 kg, 12 rounds with 30 s, resulted in significant increase in testosterone and growth hormone level in the circulation (Budnar et al., 2014).

Explosive training can increase the performance of short-, middle-, and long-distant runners, and drop- and countermovement-jump trainings can benefit basketball, volleyball, and soccer players (Ramirez-Campillo et al., 2014). After 4 weeks training on knee extensors using unilateral explosive isometric contraction, the maximal voluntary force increased significantly along with muscle-tendon stiffness, which suggests that peripheral neuronal and mechanical adaptation has taken place (Tillin et al., 2012).

Explosive training, similarly to the periodization of other abilities, is generally done using either the block or unidirectional model. It appears that in most cyclic events, such as cycling, kayak, running, cross-country skiing, swimming, and many team sports, block periodization is superior to the unidirectional training model for performance enhancement (Issurin, 2016).

12.2.8 Macro-Cycles for Speed Development

A recent study tested the effects of 40 days of training using 10 30 s maximal runs and 10 aerobic, moderate intensity runs. No significant differences were found for running economy or VO_2max between training protocols.

However, high intensity training improved short-term performance compared to aerobic training. The analysis of muscle biopsy samples showed similar results in the expression of Na^+, K^+-ATPase $\alpha 1$, $\alpha 2$, and $\beta 1$, Na^+/H^+ exchanger, sarcoplasmic reticulum by Ca^{2+} ATPases, actin, and calcium/calmodulin-dependent protein kinase (Skovgaard et al., 2018).

Seven weeks of speed interval training (Wingate protocol) resulted in increases in maximal short-term power output, VO_2max and in the enzyme activities of hexokinase, phosphofructokinase, citrate synthase, succinate dehydrogenase, and malate dehydrogenase (MacDougall et al., 1998).

In another study when subjects performed sprint interval exercise (four to eight bouts of cycling for 30 s, nine times over a duration of 3 weeks), results revealed massive increases in protein synthesis of structural and glycolytic proteins (Shankaran et al., 2016) and mitochondrial biogenesis in the skeletal muscle of the subjects (Scalzo et al., 2014). However, the increased levels of mitochondrial biogenesis could be a compensatory mechanism, since this exercise intensity leads to inhibition of aconitase, resulting in accumulation of citrate to preserve redox status (Larsen et al., 2016). Six weeks of cycling at low volume, high intensity training on untrained individuals did not result in increased cross-sectional area of working muscle, but led to an increase in the number of activated and differentiating satellite cells (Joanisse et al., 2015). This adaptive response could be important for muscle remodeling and could be associated with the need for enhanced protein synthesis. This high intensity, interval training leads to an increase in protein content of the sugar transporter GLUT4, which means that sprint exercise could be an important tool to fight against diabetes (Bradley et al., 2014). In another study, healthy subjects performed sprint training for 5 weeks, 3 days per week, with 1-min exercise bouts, which resulted in improvements in muscle Na(+), K(+)-ATPase content, and CrP resynthesis (Edge et al., 2013). Because sprint exercise stimulates a significant production of lactate, it is not surprising that sprint training is an efficient method to increase the lactate transfer capacity. It is clear that sprint exercise training results in a wide range of adaptive responses in skeletal muscle metabolism.

12.2.9 Macro-Cycles for Health Promotion and Rehabilitation

Most macro-cycles aim to increase performance in a specific sport. However, because elite sport is often associated with sport injuries, there are macro-cycles for rehabilitation and recreational activities, thus stressing health promotion. Macro-cycles for rehabilitation aim to establish the most effective programs to reacquire optimal physiological function after injury.

Most studies in rehabilitation science describe the functional effects of the intervention used in the programs, but the physiological consequences are not well described. In general, functional training targets injured muscle(s). In some diseases or during the aging process, larger muscle groups are often targeted for functional training using vibration or circuit training. Functional training is used not only in regeneration but also in top sports. In this case the functional exercise is generally done at standing position and involves multijoint movements and focusing on core muscles, joint stabilizers such as hip, torso, and posterior shoulder muscles. Functional training generally avoids large machines and loads; it uses many repetitions and trains all of the muscle around the joint, thus avoiding imbalance. Contrary to functional training in physiotherapy, in the preparation of athletes, functional training focuses on movements rather than a simple muscle group. However, it must be pointed out that the exercises that were recently termed functional exercise have been used since the preparation of ancient Greek athletes, and functional exercise has thus always been in training programs, albeit under different names.

Most rehabilitation programs use progressive loading, starting from light loads postinjury or surgery; the load is progressively increased until strength and function return to preinjury levels. Indeed, at the beginning of the post-operative period, or the beginning of the rehabilitation program, it is difficult to assess the intensity. Therefore, attempts have been made to rationalize the rehabilitation programs. There are some reports on rehabilitation after anterior cruciate ligament reconstruction (ACLR) (Horschig et al., 2014). Daily Adjustable Progressive Resistance Exercise was developed, which is an interactive protocol to determine objectively either the optimal time to increase resistance or the optimal amount of weight to increase the resistance during a resistance exercise. This method was further developed to become Autoregulatory Progressive Resistance Exercise, which introduced training cycles aimed at improving hypertrophy, strength, and power regimes of conditioning. The main criteria to bring about successful early phase improvement include restoring the active range of motion of the knee by increasing quadriceps strength (both with isometric and closed chain kinetic exercises, such as mini-squats), restoring patellar mobility, diminishing swelling and pain, and restoring safe independent ambulation (Horschig et al., 2014).

To manage posterior tightness, an 8-week stretching program for athletes using overhead motions (volleyball, tennis, swimming) has been shown to improve significantly the impaired range of motion, internal rotation, and horizontal adduction of the shoulder (Chepeha et al., 2018). Taping is also an important tool for rehabilitation of the shoulder. A recent report evaluated the effectiveness of traditional rigid and kinesio taping; the results revealed that kinesio tape is superior for glenohumeral internal rotation (rigid tape was detrimental to the injury) and range of motion (Gulpinar et al., 2017). Abdominal core activation, using calm exercise, with the dominant leg in a side prone position, hip abduction in side lying, and prone hip extension have also proven to be successful. EMG signals revealed that this abdominal core activation enhanced the therapeutic effects of hip-strengthening exercises (Chan et al., 2017).

The number of reports on the physiological and biochemical impact of periodization on health promotion and a possible treatment of various diseases is very significant. The main goals of the macro-cycle for patients suffering from different diseases are to regain a healthy status, to decrease the deleterious consequences of the disease, and to enhance the effectiveness of clinical treatments. When obese, adolescents (15–18 years old) were treated with aerobic training, or aerobic training plus strength training with linear periodization, or aerobic training with daily undulating periodization, the results revealed that each group reduced body mass and fat mass. Aerobic training was effective in the short term, while both linear and daily undulating periodization improved lipid profile, insulin sensitivity, and adiponectin concentration, showing the importance of muscle mass and function to ameliorate the metabolic syndrome and type 2 diabetes (Inoue et al., 2015). Diabetes is associated with low-grade inflammation and even a single bout of HIIT has proven to be effective in reducing the levels of Toll-like receptor 2 surface protein expression on both classical and CD16+ monocytes. Hence, HIIT has antiinflammatory effects (Durrer et al., 2017). Therefore, it is not surprising that the chronic effects of HIIT (12 weeks of training, 3 times/week, with 10×1-min cycling with about 90% of maximal heart rate) were powerful enough to decrease the levels of glycated hemoglobin, body fat, lean body mass, and blood pressure (Francois et al., 2017). Glycation of hemoglobin takes place amid normal levels of blood glucose. However, the amount significantly increases with diabetes mellitus, and this is generally used as a marker of long-term serum glucose regulation.

Testicular cancer survivors have an increased risk of treatment-related cardiovascular disease, which may limit their overall survival. Twelve weeks of supervised HIIT improved cardiorespiratory fitness and multiple pathways leading to cardiovascular risk, and decreased markers of mortality in cancer survivors (Adams et al., 2017). It is known that both moderate and high intensity regular exercise improve cardiovascular fitness, and this finding suggests that 4×4 HIIT, three times per week for at least 12 weeks, is a powerful form of exercise, which enhances vascular function (Ramos et al., 2015). However, when patients with left ventricular ejection fraction $\leq 35\%$ were asked to perform moderate or HIIT exercise for 12 weeks, although both training improved VO_2max, no difference was found in left ventricular remodeling (Ellingsen et al., 2017). Overall, the review of these related papers suggest that short interval HIIT is beneficial for coronary heart disease patients with lower aerobic fitness, and medium and/or long interval HIIT protocols may be beneficial for patients with higher aerobic fitness (Ribeiro et al., 2017).

The macro-cycles used by patients to improve their physiological functions generally are progressive, linear- or daily-undulating forms of periodization. The efficiency of the training program depends on many factors such as the state of the disease, the seriousness of the disease, age, type of applied exercise program, personal adaptability, and a list of other factors.

12.3 THE STRUCTURE AND PHYSIOLOGY OF THE PREPARATIONAL PERIOD

Goal-oriented training sessions, micro-cycles, and macro-cycles are like blocks, walls, and rooms that, along with windows and doors, build up a house, constructing periods of the year periodization. Periods are not the adaptation units of the preparation; the lengths of these units are determined by the competition structure of the year given by the event-associated national and international sport federations. Naturally, in some sports, competitors can select from the possible competitions and reduce the length of competitive period and increase the length of the preparation period. In 2017, for example, Roger Federer skipped the clay season and very successfully increased the length of recovery and preparation period for the grass season. On the other hand, Messi or C. Ronaldo had a very short preparation period, but game periodization will be discussed in separate section.

The preparational period traditionally meant a time frame to general increase in endurance and strength using high volume and low intensity exercises to optimize abilities for the competitive period to achieve maximal performance, in the case of elite sport. However, as a result of significant increase in the competitive season(s) in most sports, the goals and structure of the preparation period have also changed. These days, only a very few sports have the luxury of a preparation period that is long enough to develop, and increase the level of the crucial, crude abilities that determine the athlete's level of performance during the competitive period(s).

However, regardless on the length of the preparation period, it is generally true that in this period the training sessions in the micro- and macrocycle(s) have high volume and low intensity. However, high jumpers in track and field do not have to run 30 km per week, and increase muscle hypertrophy and neglect high intensity training in this period, despite the fact that in the preparation period, we generally use high volume and low intensity. The significant training load in this period increases the chance of overtraining, so care is required to avoid it.

In elite sport it is important to focus mainly on that ability which based on the objective measurements most significantly limits the performance of the athletes. If we take triathlon as an example, those who are recruited into this sport without significant background in swimming, besides enhancing aerobic performance, targeting to improve their technique in this event has a beneficial impact on the performance. Moreover, if speed is the greatest limitation of the performance, it is better to forget trainings with low intensity and high duration, and to increase the intensity and the speed even in the periodization season. Therefore, in modern periodization the difference between the preparational and competitive periods is not as great as before, and the loading is quite similar thorough the year; however, different goals in different periods are still present. In this period athletes are not competing; they are not under pressure to achieve maximal performance.

12.4 THE STRUCTURE AND PHYSIOLOGY OF THE COMPETITIVE PERIOD

In the competition period, the main goal is to achieve maximal or submaximal performance. Depending on the goal set, some few main competitions are selected by individual athletes. The number varies by sport and competitors. In general, in track and field and swimming 2–4 main completions in a year can be selected, but this also depends on the event and competitor. Marathon runners have fewer main competitions than 100 m runners do, while Hungarian Olympic swimmer Katinka Hosszu has many swimming competition with great results in a year. In this period the training volume is low and the intensity is high. High intensity is a must in the competition period because of the close link between intensity and performance; due to this, competition offers the best training. Therefore, one might require some competitions to peak the performance, but as with everything, this is also event and competitor dependent. Because of the increased intensity, the loading provides different challenges to the body which can easily lead to injury. Hence, special attention is required at the transition between preparation and competitive periods.

During the competitive period, functional physiological markers such as maximal strength, explosive strength, VO_2max, anaerobic threshold, blood lactate peak concentration, economy of the cyclic movements, maximal speed, acceleration, etc. must be optimal. Because of the increases in the intensity of the exercise and because of the relationship between intensity and anabolic hormone secretion, athletes' bodies more often experience greater amounts of IGF-1, testosterone, and growth hormone (Papacosta et al., 2013; Tourinho Filho et al., 2017). Interestingly, an increased level of testosterone can even beneficially affect mood (Pope et al., 2010), cognitive function (Jia et al., 2013), and cardiovascular capacity (Kelly and Jones, 2014); therefore one could suggest that greater testosterone levels due to high intensity would enhance recovery and performance.

The preparation of athletes during tapering to competition is more sport-specific. Peaking results in increases in intensity and decline in duration of training sessions. Moderate intensity exercise with long duration can improve the economy of running; high intensity exercise can more efficiently increase VO_2max but decrease the efficiency of running (Munoz et al., 2015). In elite cross-country skiers and biathletes, the reduction in training volume during peaking is around 32% compared to the preparation period (Tonnessen et al., 2014). The importance of training intensity, especially in the peaking and tapering phases, can be well understood by the study of Hickson et al. (1985). They trained 12 subjects for 10 weeks, but then reduced the intensity by one-third or two-thirds, and then continued the training for additional 15 weeks (Hickson et al., 1985). After 10 weeks the increase in VO_2max was between 11% and 20%, but after the reduction of exercise intensity the VO_2max was not maintained and reduced in a greater extent in the group that had reduced the intensity by two-thirds. The short-term endurance up to 5 min was still maintained in the group that had decreased the intensity by one-third, but not the long-term endurance, while both short- and long-term endurance decreased in the group that decreased the intensity by two-thirds. Echocardiographic data revealed that 10 weeks of training increased the left ventricular mass by 15%; this was lost after the reduction of exercise intensity even with the maintained duration in both groups (Hickson et al., 1985). This study evidences the importance of intensity to maintain important components of endurance capacity.

When the ordinary training duration (approximately 45 km/week running) was changed to training with frequent high-intensity sessions each consisting of 8–12 30-s sprint runs separated by 3 min of rest (5.7 ± 0.1 km/week) with additional 9.9 ± 0.3 km/week at low running speed for 4 weeks, results revealed improved running economy at submaximal running (Iaia et al., 2009). In this study, high intensity interval training reduced energy expenditure of running, but did not change oxidative capacity of skeletal muscle or capillarization in well-trained runners.

When tapering of middle-distance runners was studied either by high intensity and low volume, or low intensity with moderate volume or rest, the results pointed out that only high intensity low volume tapering increased the running time to fatigue, citrate synthase activity, and blood volume (Shepley et al., 1992). A similar study was done by Mujika and co-workers, who studied the 6 days tapering of middle-distance runners and concluded that high intensity training with low volume was most effective, and well-trained middle-distance runners can reduce their usual training volume by 75% in the last week before the competition (Mujika et al., 2000). High intensity tapering out-performed low intensity tapering at blood volume and testosterone levels, and the latter could be important to facilitate recovery as well.

During preparation, most of the training time is spent in the low intensity zone, which during the taper is reduced, and the training time at the higher intensity period is increased slightly.

12.5 THE STRUCTURE OF TRANSITION AND OFF-SEASON PERIODS

The goal of transition is to facilitate psychological rest, relaxation, and biological regeneration, as well as to maintain an acceptable level of general physical preparation. The length of this period is quite short due to the busy program of most athletes. During the off-season, the main goal is to have complete regeneration (psychological and physical), and to prepare for the next period.

12.6 THE STRUCTURE AND GOALS OF A 1-YEAR CYCLE OF PERIODIZATION

In most sports, coaches use a 1-year periodization cycle which contains training sessions, micro- and macro-cycles, preparation, competitive, transition, and off-season periods, as discussed earlier. Similarly to each component of 1-year periodization, the 1-year plan is also goal driven. In top sport the obvious goal is to reach maximal performance; however, goal setting at young, developing athletes is very complex due to biological changes, which is part of the maturation process.

12.7 LONG-TERM ATHLETE DEVELOPMENT

The goals of training change gradually, from children to professional athletes, from beginners to Olympic champions. In the beginning, for children, the most important points are to provide fun and enjoyment, and to create a strong bond to the given sport. The maturation of the body determines the most successful periods to enhance strength, endurance, speed, flexibility, and a technical repertoire. If one accepts that strength development is significantly influenced by the maturation of bone, muscle, and the hormonal systems, then efficient strength training cannot be successfully applied before the required pubertal changes occur. Large individual differences in maturation rates can be seen that are not related to the chronological age of adolescents, and this demands special attention from coaches. For example, height increases dramatically around 13–16 years of age in boys, when the production of growth hormone shows a significant elevation (Fig. 12.5), but the spurt is not equivalent among all adolescents.

This change might determine not only which sport the child selects, but also what success the child might enjoy. Nonetheless, the periodization would seldom be similar for a group of adolescents. In boys the free testosterone

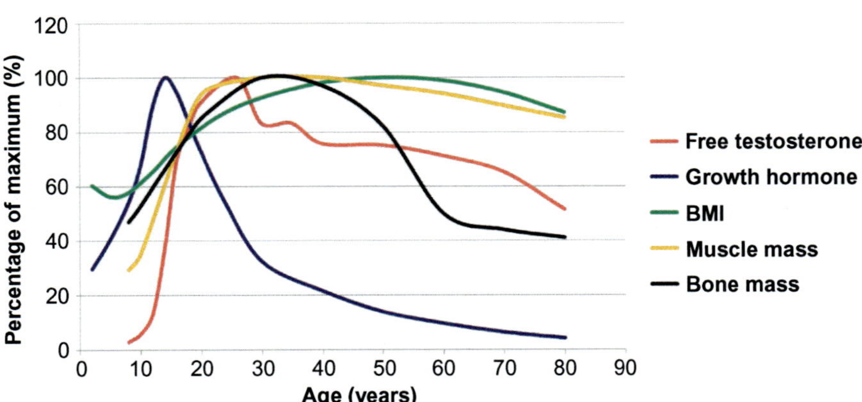

Fig. 12.5 Maturation hormonal and skeletal systems. The production of anabolic hormones (growth hormone and testosterone) and the related development of muscle and bone systems strongly affect the optimal time period of the development of conditional abilities.

levels also rise, providing the anabolic background for bone and muscle development. The general belief is that resistance training should not be used before adolescence. One of the potential age-associated risks of resistance training for prepubertal children is the damage to growth plates, which could have long-term consequences. Moreover, according to recent reports, the overall injury to prepubertal children who are involved in resistance training is relatively low, and the injury is probably due to poor training techniques and coaching, and the fact that heavy weights are seldom used (Faigenbaum and Myer, 2010). It has been suggested that well-designed programs of resistance training, before adolescence, could even decrease the risk of sport injuries by increasing muscular strength (Myers et al., 2017). Interestingly, meta-analysis revealed no age-associated difference to the adaptive response to resistance training in age ranging between 6 and 18 years (Lesinski et al., 2016). A growing body of evidence suggests that well-designed resistance training for preadolescents could positively affect muscular strength and endurance, and increase normal growth and maturation (Behm et al., 2017). Reactive training, also called plyometric training, which can include jumping, hopping, and bounding exercises and throws, performed quickly and explosively is especially effective in preadolescents and reportedly improves speed, agility, and performance (Johnson et al., 2011). On the other hand, it must be noted that greater levels of growth hormone, testosterone, and well-developed musculature after adolescence provide a better background to develop strength, explosive strength, and hypertrophy, and the prime time to develop these abilities may be between 18 and 30 years.

A greater increase in growth hormone (GH) levels has impact on endurance capacity as well. GH is linked to increases in ventricular size and ejection fraction of the heart (Borer, 1995). In adolescent boys, positive correlations were found between the mean overnight levels of GH and peak oxygen uptake (Eliakim et al., 1998), which could be related to increased heart and muscle mass. When three 30-min aerobic exercises per week were given for 12 weeks to 11–13-year-old boys, the VO_2max increased from 44.7 (5.8) to 47.6 (6.4) mL/kg/min only in the exercise group (Rowland and Boyajian, 1995). Young athletes have higher heart rates, better heart rate recovery, and lower stroke volume and cardiac output than adults, but this is the only meaningful difference in their response to endurance exercise (Fig. 12.6).

Preadolescent endurance training is important to achieve peak endurance capacity at adulthood. However, children have been suspected of having a diminished capability of responding to endurance training with improvements of maximal oxygen uptake compared to adults. Hence, training with

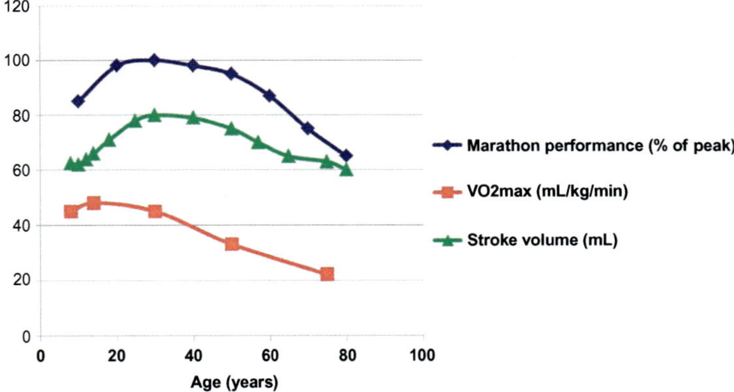

Fig. 12.6 The maturation of cardiovascular system effects endurance performance. Although endurance capacity is a complex ability, the age-dependent changes in heart function and VO_2max influence the performance in endurance and the trainability of endurance.

the same load would result in a smaller degree of improvement in aerobic capacity in preadolescents than in adults.

In terms of speed training, preadolescent training, especially with a focus on technique, is very important and effective. The development of the nervous system allows one to master sport techniques, the levels of which can correlate well with the speed of a particular sport (Fig. 12.7).

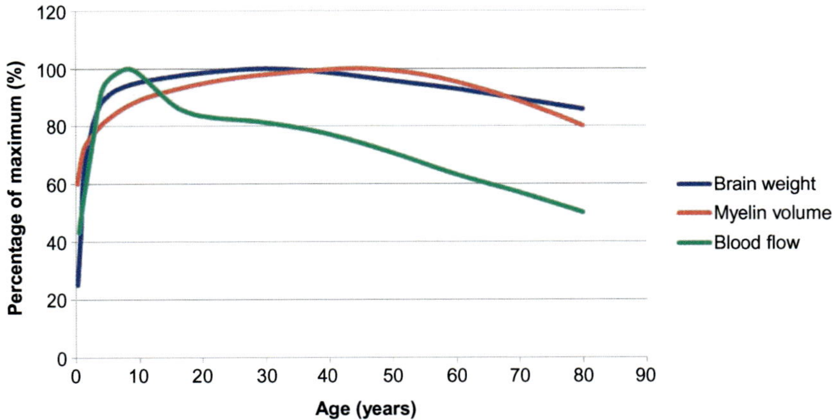

Fig. 12.7 Age-related changes in the neuro-system. This figure illustrates the age-associated changes in weight of the brain, blood flow in the brain, and myelination of neurons. These changes strongly influence the time period in which complex motor techniques and speed can be developed.

However, because speed is dependent on explosive strength, which reaches optimum levels in athletes generally older than 18 years, due to the maturation of muscle and the hormonal system, speed could be efficiently improved in adulthood by explosive strength training. Indeed, due to the strong dependence on sport technique and explosive strength, the prime age to achieve top performance in events like javelin, discus, or hammer throwing is often over 25 years. But in sports where speed and tactics are very important, like fencing, hand ball, or soccer, athletes older than 30 years are often found to have the greatest success.

Flexibility is a conditional ability that can be developed at the earliest time, even from the age of 5. Then the focus can be the wide range of development of coordinational abilities, therefore basic sport techniques could be efficiently learned at a young age (Baar, 2014; Baggish et al., 2014; Bangsbo et al., 2006; Bartolomei et al., 2014; Behm et al., 2017; Bishop et al., 2014; Borer, 1995; Bradley et al., 2014).

Although the response to aerobic endurance training is slightly suppressed in preadolescence, this training must be done to provide the necessary cardiovascular background for top sport. Resistance training can be also started at preadolescence but the focus must be on explosive strength, while the maximal strength related part of resistance training should be delayed until after maturation.

12.8 CHARACTERISTICS OF BALL GAME PERIODIZATION

The periodization of ball games is unique due to the very long competitive period(s) (12–35 per week) and the number of players in different positions requiring different conditions and skills. The periodization of the team will undoubtedly differ from the periodization of individual athletes. Theoretically individual periodization also can be done, but usually not every player in the team has an individual program. However, players who exhibit very different conditions or skills from the mean of the team deserve individual preparational plans. In addition, key or star players will have individualized training plans. Moreover, injured players will be guided to recovery programs. Coaches must handle the team as a goal-oriented unit, but at the same time they must improve the quality of each individual player. In other words, the periodization plans of the individual athletes must add up to and equal the periodization of the team. This fact leads to a very difficult and complex job for the coach and team management. The other challenging problem is the very long competitive season, which can last up to 40 weeks (e.g., 80–100 games). It is impossible to play every game at 100% of peak performance (which cannot really be accurately measured in team sports). Therefore, it is crucial to attain peak performance for the most important game such as league championships or Olympic finals.

The given length of preparational and competitive periods makes it extremely difficult to exercise great differences between these periods. One cannot avoid targeting to improve performance in strength and endurance, even in the competitive period(s) in order to provide a physiological basis for technical and tactical tasks. In order to optimize the performance of teams, a match difficulty index (MDI) has been created (Robertson and Joyce, 2015), which evaluates individual weightings of fixed and dynamic factors based on their influence in determining the difficulty of matches. Among the fixed factors, one can find the ranking of the opponent team in the early season, the location of matches (home or away), and the number of days between matches. The dynamic factors cover the current position and point number of the opposition, the current condition of the home team, accumulation of fatigue in both teams, the weather, the number of spectators, the noise factor within certain stadia, the visiting team's track record against a certain opponent, etc. Creation of the MDI has helped to define those matches in which the team performance must be maximal or close to maximal, so special attention and peaking is important.

Tennis has a very long competitive season and the best players play up to 70 games per year. Due to this, and the need for explosive force (serving), it is not surprising that the players continue resistance training throughout the year. It has been shown that greater strength gains, due to periodized resistance training (4–6 RM 2–3 times per week up to 9 months), contribute to large improvements in jump height, and ball velocity for the serve, forehand, and backhand strokes in female tennis players (Kraemer et al., 2003).

When tapering in professional English soccer players was examined, it turned out that decreasing training load during taper weeks by reducing training duration and frequency, but maintaining intensity, was associated with an increase in physical activity during matches (Fessi et al., 2016).

Peaking at ball sports for the Olympic Games or World or European championships is very tricky. These international events last 2–3 weeks and the finalists play up to seven or eight games in that time frame. In these events, performance of the team must be optimal and few teams can afford to increase performance gradually to try to peak for only the final match. These events need a special periodization program that takes into account the fact that most of the coaches have a short time frame to prepare for these events, perhaps due to the busy schedule of players, to create the most effective team possible. One of the main goals is to create a powerful team, with effective tactical repertoires.

12.9 "R" EXTRA

VO_2max is one of the best available markers of viability of humans, and the level attained is a limiting factor in many sports. Therefore, the trainability of VO_2max is a key issue. It has been suggested that the trainability of VO_2max is dependent, at least in part, on genetics. Timmons and co-workers (Keller et al., 2011) developed a system called training-responsive transcriptome, to investigate the genomic effects of aerobic trainability. They identified three DNA sequences: the RUNX1, SOX9, and PAX3 transcription factor binding sites in training-response transcriptome (Keller et al., 2011). It was observed that aerobic training responders and nonresponders show quite different gene regulation patterns, and responders express genes that are involved in angiogenesis, tissue repair, and development.

Few sports have the luxury of being able to avoid concurrent training loads, which means that in one micro- or macro-cycle, different abilities are developed in most sports. However, the interference between different conditional abilities could affect the performance positively or negatively. When strength and endurance training are given to subjects, alone or both types of training together with the same training load, the results revealed that VO_2max increased in endurance and endurance and strength training group and naturally did change in strength training group. On the other hand, the strength increased to a greater degree in the strength training group, and the continuous endurance training attenuated the increased in strength in the group that done both types of training simultaneously (Hickson, 1980). Therefore, these results suggest that concurrent training (endurance and muscle mass-based strength training) attenuates the development of strength but does not alter the development of VO_2max. However, data suggest that strength and endurance increase concomitantly up to a point, and above the intensity 80% VO_2max, endurance exercise prevents the increase in muscle mass and strength that occur with strength training (Baar, 2014). What could be the molecular mechanisms by which endurance negatively interferes with strength training? It cannot be inter- or intramuscular adaptation, as any improvement could be responsible for the early phase increase in strength development. Rather, it could be interference in cellular signaling pathways. One possible explanation by which high intensity endurance training suppresses the gain in strength after a period of time could be through the AMPK mediated inhibition of mTOR signaling. AMPK is readily activated by high intensity exercise. It has been shown that metabolic stress results in enhanced phosphorylation of TSC2, which is a tumor suppressor of AMPK (Inoki et al., 2003), and TSC2 can inhibit the phosphorylation of ribosomal S6 kinase (S6K) and eukaryotic initiation factor 4E binding protein-1 (4EBP1), which are mTOR substrates (Inoki et al., 2003). Indeed, it has been shown that increased activity of AMPK can attenuate hypertrophy in aged plantaris caused by overload (Thomson and Gordon, 2005). Interestingly, when repeated blood flow restriction was used on rats to cause hypertrophy, mTOR was activated, while AMPK phosphorylation was unchanged (Nakajima et al., 2016). Because the effectiveness of strength training is dependent on the intensity of following endurance exercise in concurrent training, it is advised to use endurance exercise of intensity lower than 70% to optimize the effects of strength training or provide time to complete regeneration after high intensity exercise before strength training.

When the effects of endurance training on muscle capillarization were studied, it turned out that 8 weeks of training on a bicycle ergometer for an average of 40 min/day, four times a week at a work load requiring 80% of the maximal oxygen uptake, resulted in increased capillarization in type I and type IIa and type IIb fibers (Andersen and Henriksson, 1977). An increased level of capillarization is important to enhance VO_2max and endurance.

12.10 SUMMARY

Periodization involves goal-driven planning, divided into single training sessions, micro- and macro-cycles, preparational, competition, transition, and off-season periods. The goals of periodization vary according to the level of physical fitness, whether the athletes are in competitive or recreation activity, age among other factors. The goals of professional preparation focus on the achievement of maximal performance, while in athletes' long-term development, the goals are strongly determined by physiological development and maturation. About 4–12 weeks, a micro-cycle is necessary to cause adaptation that leads to increased performance. The effects of exercise load are specific and largely dependent on intensity, which influences the type of fatigue, cellular interactions, and recovery periods. The pattern of loading can be nonlinear periodization, including the undulating model or block periodization. At the top level of sports, periodization during the preparational period, the ability which mostly limits the performance, must be most extensively improved and in this period the duration of the training load is emphasized. On the other hand, during the competitive period the intensity increases, leading to enhanced performance.

TEST QUESTIONS

1. What are the components of 1-year periodization?
2. What factors determine goal settings in periodization?
3. What kind of loading patterns do you know?
4. What are the physiological effects of a micro-cycle that aims to improve aerobic endurance?
5. What are the physiological effects of a micro-cycle that aims to improve explosive strength?
6. What are the physiological effects of a micro-cycle that aims to improve speed?
7. How would you prepare long-term development exercise program for children?

BIBLIOGRAPHY

Abe, T., Kitaoka, Y., Kikuchi, D.M., Takeda, K., Numata, O., Takemasa, T., 2015. High-intensity interval training-induced metabolic adaptation coupled with an increase in Hif-1alpha and glycolytic protein expression. J. Appl. Physiol. 119, 1297–1302.

Adams, S.C., DeLorey, D.S., Davenport, M.H., Stickland, M.K., Fairey, A.S., North, S., Szczotka, A., Courneya, K.S., 2017. Effects of high-intensity aerobic interval training on cardiovascular disease risk in testicular cancer survivors: a phase 2 randomized controlled trial. Cancer 123, 4057–4065.

Afzalpour, M.E., Chadorneshin, H.T., Foadoddini, M., Eivari, H.A., 2015. Comparing interval and continuous exercise training regimens on neurotrophic factors in rat brain. Physiol. Behav. 147, 78–83.

Andersen, P., Henriksson, J., 1977. Capillary supply of the quadriceps femoris muscle of man: adaptive response to exercise. J. Physiol. 270, 677–690.

Baar, K., 2014. Using molecular biology to maximize concurrent training. Sports Med. 44 (Suppl. 2), S117–125.

Baggish, A.L., Park, J., Min, P.K., Isaacs, S., Parker, B.A., Thompson, P.D., Troyanos, C., D'Hemecourt, P., Dyer, S., Thiel, M., Hale, A., Chan, S.Y., 2014. Rapid upregulation and clearance of distinct circulating microRNAs after prolonged aerobic exercise. J. Appl. Physiol. 116, 522–531.

Bangsbo, J., Mohr, M., Krustrup, P., 2006. Physical and metabolic demands of training and match-play in the elite football player. J. Sports Sci. 24, 665–674.

Bartolomei, S., Hoffman, J.R., Merni, F., Stout, J.R., 2014. A comparison of traditional and block periodized strength training programs in trained athletes. J. Strength Cond. Res. 28, 990–997.

Behm, D.G., Young, J.D., Whitten, J.H.D., Reid, J.C., Quigley, P.J., Low, J., Li, Y., Lima, C.D., Hodgson, D.D., Chaouachi, A., Prieske, O., Granacher, U., 2017. Effectiveness of traditional strength vs. power training on muscle strength, power and speed with youth: a systematic review and meta-analysis. Front. Physiol. 8, 423.

Bishop, D.J., Granata, C., Eynon, N., 2014. Can we optimise the exercise training prescription to maximise improvements in mitochondria function and content? Biochim. Biophys. Acta 1840, 1266–1275.

Borer, K.T., 1995. The effects of exercise on growth. Sports Med. 20, 375–397.

Bradley, H., Shaw, C.S., Worthington, P.L., Shepherd, S.O., Cocks, M., Wagenmakers, A.J., 2014. Quantitative immunofluorescence microscopy of subcellular GLUT4 distribution in human skeletal muscle: effects of endurance and sprint interval training. Phys. Rep. 2.

Brooks, G.A., 1986. Lactate production under fully aerobic conditions: the lactate shuttle during rest and exercise. Fed. Proc. 45, 2924–2929.

Budnar Jr., R.G., Duplanty, A.A., Hill, D.W., McFarlin, B.K., Vingren, J.L., 2014. The acute hormonal response to the kettlebell swing exercise. J. Strength Cond. Res. 28, 2793–2800.

Casadio, J.R., Storey, A.G., Merien, F., Kilding, A.E., Cotter, J.D., Laursen, P.B., 2017. Acute effects of heated resistance exercise in female and male power athletes. Eur. J. Appl. Physiol. 117, 1965–1976.

Chan, M.K., Chow, K.W., Lai, A.Y., Mak, N.K., Sze, J.C., Tsang, S.M., 2017. The effects of therapeutic hip exercise with abdominal core activation on recruitment of the hip muscles. BMC Musculoskelet. Disord. 18, 313.

Chepeha, J.C., Magee, D.J., Bouliane, M., Sheps, D., Beaupre, L., 2018. Effectiveness of a posterior shoulder stretching program on university-level overhead athletes: randomized controlled trial. Clin. J. Sport Med. 28, 146–152.

Choi, D., Cole, K.J., Goodpaster, B.H., Fink, W.J., Costill, D.L., 1994. Effect of passive and active recovery on the resynthesis of muscle glycogen. Med. Sci. Sports Exerc. 26, 992–996.

Coetsee, C., Terblanche, E., 2017. Cerebral oxygenation during cortical activation: the differential influence of three exercise training modalities. A randomized controlled trial. Eur. J. Appl. Physiol. 117, 1617–1627.

Degoutte, F., Jouanel, P., Filaire, E., 2003. Energy demands during a judo match and recovery. Br. J. Sports Med. 37, 245–249.

Di Blasio, A, Izzicupo, P, Tacconi, L, Di Santo, S, Leogrande, M, Bucci, I, Ripari, P, Di Baldassarre, A & Napolitano, G. (2016). Acute and delayed effects of high intensity interval resistance training organization on cortisol and testosterone production. J. Sports Med. Phys. Fitness 56, 192–199.

Durrer, C., Francois, M., Neudorf, H., Little, J.P., 2017. Acute high-intensity interval exercise reduces human monocyte Toll-like receptor 2 expression in type 2 diabetes. Am. J. Physiol. Regul. Integr. Comp. Physiol. 312, R529–R538.

Edge, J., Eynon, N., McKenna, M.J., Goodman, C.A., Harris, R.C., Bishop, D.J., 2013. Altering the rest interval during high-intensity interval training does not affect muscle or performance adaptations. Exp. Physiol. 98, 481–490.

Egan-Shuttler, J.D., Edmonds, R., Eddy, C., O'Neill, V., Ives, S.J., 2017. The effect of concurrent plyometric training versus submaximal aerobic cycling on rowing economy, peak power, and performance in male high school rowers. Sports Med. Open 3, 7.

Eliakim, A., Brasel, J.A., Barstow, T.J., Mohan, S., Cooper, D.M., 1998. Peak oxygen uptake, muscle volume, and the growth hormone-insulin-like growth factor-I axis in adolescent males. Med. Sci. Sports Exerc. 30, 512–517.

Ellingsen, O., Halle, M., Conraads, V., Stoylen, A., Dalen, H., Delagardelle, C., Larsen, A.I., Hole, T., Mezzani, A., Van Craenenbroeck, E.M., Videm, V., Beckers, P., Christle, J.W., Winzer, E., Mangner, N., Woitek, F., Hollriegel, R., Pressler, A., Monk-Hansen, T., Snoer, M., Feiereisen, P., Valborgland, T., Kjekshus, J., Hambrecht, R., Gielen, S., Karlsen, T., Prescott, E., Linke, A., Group, SHFS, 2017. High-intensity interval training in patients with heart failure with reduced ejection fraction. Circulation 135, 839–849.

Esteve-Lanao, J., San Juan, A.F., Earnest, C.P., Foster, C., Lucia, A., 2005. How do endurance runners actually train? Relationship with competition performance. Med. Sci. Sports Exerc. 37, 496–504.

Esteve-Lanao, J., Rhea, M.R., Fleck, S.J., Lucia, A., 2008. Running-specific, periodized strength training attenuates loss of stride length during intense endurance running. J. Strength Cond. Res. 22, 1176–1183.

Evans, M., Cogan, K.E., Egan, B., 2017. Metabolism of ketone bodies during exercise and training: physiological basis for exogenous supplementation. J. Physiol. 595, 2857–2871.

Faigenbaum, A.D., Myer, G.D., 2010. Resistance training among young athletes: safety, efficacy and injury prevention effects. Br. J. Sports Med. 44, 56–63.

Fessi, M.S., Zarrouk, N., Di Salvo, V., Filetti, C., Barker, A.R., Moalla, W., 2016. Effects of tapering on physical match activities in professional soccer players. J. Sports Sci. 34, 2189–2194.

Figueiredo, V.C., Caldow, M.K., Massie, V., Markworth, J.F., Cameron-Smith, D., Blazevich, A.J., 2015. Ribosome biogenesis adaptation in resistance training-induced human skeletal muscle hypertrophy. Am. J. Physiol. Endocrinol. Metab. 309, E72–83.

Fournier, P.E., Stalder, J., Mermillod, B., Chantraine, A., 1997. Effects of a 110 kilometers ultra-marathon race on plasma hormone levels. Int. J. Sports Med. 18, 252–256.

Francois, M.E., Durrer, C., Pistawka, K.J., Halperin, F.A., Chang, C., Little, J.P., 2017. Combined interval training and post-exercise nutrition in type 2 diabetes: a randomized control trial. Front. Physiol. 8, 528.

Frost, D.M., Bronson, S., Cronin, J.B., Newton, R.U., 2016. Changes in maximal strength, velocity, and power after 8 weeks of training with pneumatic or free weight resistance. J. Strength Cond. Res. 30, 934–944.

Gavin, T.P., Drew, J.L., Kubik, C.J., Pofahl, W.E., Hickner, R.C., 2007. Acute resistance exercise increases skeletal muscle angiogenic growth factor expression. Acta Physiol. (Oxf.) 191, 139–146.

Gejl, K.D., Hvid, L.G., Frandsen, U., Jensen, K., Sahlin, K., Ortenblad, N., 2014. Muscle glycogen content modifies SR Ca^{2+} release rate in elite endurance athletes. Med. Sci. Sports Exerc. 46, 496–505.

Giovanelli, N., Taboga, P., Rejc, E., Lazzer, S., 2017. Effects of strength, explosive and plyometric training on energy cost of running in ultra-endurance athletes. Eur. J. Sport Sci. 17, 805–813.

Guillou, P., 1989. Interleukin-2 and lymphokine-activated killer cell therapy for gastrointestinal cancer. Acta Chir. Scand. Suppl. 549, 26–30.

Gulpinar, D., Ozer, S.T., Yesilyaprak, S.S., 2017. Effects of rigid and Kinesio taping on shoulder rotation motions, posterior shoulder tightness, and posture in asymptomatic overhead athletes: a randomized controlled trial. J. Sport Rehabil., 1–26.

Harber, M.P., Gallagher, P.M., Creer, A.R., Minchev, K.M., Trappe, S.W., 2004. Single muscle fiber contractile properties during a competitive season in male runners. Am. J. Physiol. Regul. Integr. Comp. Physiol. 287, R1124–1131.

Hickson, R.C., 1980. Interference of strength development by simultaneously training for strength and endurance. Eur. J. Appl. Physiol. Occup. Physiol. 45, 255–263.

Hickson, R.C., Foster, C., Pollock, M.L., Galassi, T.M., Rich, S., 1985. Reduced training intensities and loss of aerobic power, endurance, and cardiac growth. J. Appl. Physiol. 58, 492–499.

Hofmann, P., Tschakert, G., 2017. Intensity- and duration-based options to regulate endurance training. Front. Physiol. 8, 337.

Horschig, A.D., Neff, T.E., Serrano, A.J., 2014. Utilization of autoregulatory progressive resistance exercise in transitional rehabilitation periodization of a high school football-player following anterior cruciate ligament reconstruction: a case report. Int. J. Sports Phys. Ther. 9, 691–698.

Iaia, F.M., Hellsten, Y., Nielsen, J.J., Fernstrom, M., Sahlin, K., Bangsbo, J., 2009. Four weeks of speed endurance training reduces energy expenditure during exercise and maintains muscle oxidative capacity despite a reduction in training volume. J. Appl. Physiol. 106, 73–80.

Inoki, K., Zhu, T., Guan, K.L., 2003. TSC2 mediates cellular energy response to control cell growth and survival. Cell 115, 577–590.

Inoue, D.S., De Mello, M.T., Foschini, D., Lira, F.S., De Piano Ganen, A., Da Silveira Campos, R.M., De Lima Sanches, P., Silva, P.L., Corgosinho, F.C., Rossi, F.E., Tufik, S., Damaso, A.R., 2015. Linear and undulating periodized strength plus aerobic training promote similar benefits and lead to improvement of insulin resistance on obese adolescents. J. Diabetes Complicat. 29, 258–264.

Issurin, V.B., 2016. Benefits and limitations of block periodized training approaches to athletes' preparation: a review. Sports Med. 46, 329–338.

Jia, J., Kang, L., Li, S., Geng, D., Fan, P., Wang, L., Cui, H., 2013. Amelioratory effects of testosterone treatment on cognitive performance deficits induced by soluble Abeta1-42 oligomers injected into the hippocampus. Horm. Behav. 64, 477–486.

Joanisse, S., McKay, B.R., Nederveen, J.P., Scribbans, T.D., Gurd, B.J., Gillen, J.B., Gibala, M.J., Tarnopolsky, M., Parise, G., 2015. Satellite cell activity, without expansion, after non-hypertrophic stimuli. Am. J. Physiol. Regul. Integr. Comp. Physiol. 309, R1101–1111.

Johnson, B.A., Salzberg, C.L., Stevenson, D.A., 2011. A systematic review: plyometric training programs for young children. J. Strength Cond. Res. 25, 2623–2633.

Kao, S.C., Westfall, D.R., Soneson, J., Gurd, B., Hillman, C.H., 2017. Comparison of the acute effects of high-intensity interval training and continuous aerobic walking on inhibitory control. Psychophysiology 54, 1335–1345.

Keller, P., Vollaard, N.B., Gustafsson, T., Gallagher, I.J., Sundberg, C.J., Rankinen, T., Britton, S.L., Bouchard, C., Koch, L.G., Timmons, J.A., 2011. A transcriptional map of the impact of endurance exercise training on skeletal muscle phenotype. J. Appl. Physiol. 110, 46–59.

Kelly, D.M., Jones, T.H., 2014. Testosterone and cardiovascular risk in men. Front. Horm. Res. 43, 1–20.

Kilian, Y., Wehmeier, U.F., Wahl, P., Mester, J., Hilberg, T., Sperlich, B., 2016. Acute response of circulating vascular regulating microRNAs during and after high-intensity and high-volume cycling in children. Front. Physiol. 7, 92.

Kohn, T.A., Essen-Gustavsson, B., Myburgh, K.H., 2011. Specific muscle adaptations in type II fibers after high-intensity interval training of well-trained runners. Scand. J. Med. Sci. Sports 21, 765–772.

Koltai, E., Bori, Z., Chabert, C., Dubouchaud, H., Naito, H., Machida, S., Davies, K.J., Murlasits, Z., Fry, A.C., Boldogh, I., Radak, Z., 2017. SIRT1 may play a crucial role in overload-induced hypertrophy of skeletal muscle. J. Physiol. 595, 3361–3376.

Kraemer, W.J., Hakkinen, K., Newton, R.U., McCormick, M., Nindl, B.C., Volek, J.S., Gotshalk, L.A., Fleck, S.J., Campbell, W.W., Gordon, S.E., Farrell, P.A., Evans, W.J., 1998. Acute hormonal responses to heavy resistance exercise in younger and older men. Eur. J. Appl. Physiol. Occup. Physiol. 77, 206–211.

Kraemer, W.J., Hakkinen, K., Triplett-Mcbride, N.T., Fry, A.C., Koziris, L.P., Ratamess, N.A., Bauer, J.E., Volek, J.S., McConnell, T., Newton, R.U., Gordon, S.E., Cummings, D., Hauth, J., Pullo, F., Lynch, J.M., Fleck, S.J., Mazzetti, S.A., Knuttgen, H.G., 2003. Physiological changes with periodized resistance training in women tennis players. Med. Sci. Sports Exerc. 35, 157–168.

Krustrup, P., Ortenblad, N., Nielsen, J., Nybo, L., Gunnarsson, T.P., Iaia, F.M., Madsen, K., Stephens, F., Greenhaff, P., Bangsbo, J., 2011. Maximal voluntary contraction force, SR function and glycogen resynthesis during the first 72 h after a high-level competitive soccer game. Eur. J. Appl. Physiol. 111, 2987–2995.

Lake, J.P., Lauder, M.A., 2012. Kettlebell swing training improves maximal and explosive strength. J. Strength Cond. Res. 26, 2228–2233.

Larsen, F.J., Schiffer, T.A., Ortenblad, N., Zinner, C., Morales-Alamo, D., Willis, S.J., Calbet, J.A., Holmberg, H.C., Boushel, R., 2016. High-intensity sprint training inhibits mitochondrial respiration through aconitase inactivation. FASEB J. 30, 417–427.

Lauber, B., Franke, S., Taube, W., Gollhofer, A., 2017. The effects of a single bout of exercise on motor memory interference in the trained and untrained hemisphere. Neuroscience 347, 57–64.

Leiter, J.R., Peeler, J., Anderson, J.E., 2011. Exercise-induced muscle growth is muscle-specific and age-dependent. Muscle Nerve 43, 828–838.

Lesinski, M., Prieske, O., Granacher, U., 2016. Effects and dose-response relationships of resistance training on physical performance in youth athletes: a systematic review and meta-analysis. Br. J. Sports Med. 50, 781–795.

Linnamo, V., Newton, R.U., Hakkinen, K., Komi, P.V., Davie, A., McGuigan, M., Triplett-McBride, T., 2000. Neuromuscular responses to explosive and heavy resistance loading. J. Electromyogr. Kinesiol. 10, 417–424.

Little, J.P., Safdar, A., Wilkin, G.P., Tarnopolsky, M.A., Gibala, M.J., 2010. A practical model of low-volume high-intensity interval training induces mitochondrial biogenesis in human skeletal muscle: potential mechanisms. J. Physiol. 588, 1011–1022.

Loturco, I., Nakamura, F.Y., Kobal, R., Gil, S., Pivetti, B., Pereira, L.A., Roschel, H., 2016. Traditional periodization versus optimum training load applied to soccer players: effects on neuromuscular abilities. Int. J. Sports Med. 37, 1051–1059.

Lundby, C., Robach, P., 2015. Performance enhancement: what are the physiological limits? Physiology 30, 282–292.

MacDougall, J.D., Hicks, A.L., MacDonald, J.R., McKelvie, R.S., Green, H.J., Smith, K.M., 1998. Muscle performance and enzymatic adaptations to sprint interval training. J. Appl. Physiol. 84, 2138–2142.

Mangine, G.T., Hoffman, J.R., Gonzalez, A.M., Townsend, J.R., Wells, A.J., Jajtner, A.R., Beyer, K.S., Boone, C.H., Wang, R., Miramonti, A.A., LaMonica, M.B., Fukuda, D.H., Witta, E.L., Ratamess, N.A., Stout, J.R., 2017. Exercise-induced hormone elevations are related to muscle growth. J. Strength Cond. Res. 31, 45–53.

McConell, G.K., Costill, D.L., Widrick, J.J., Hickey, M.S., Tanaka, H., Gastin, P.B., 1993. Reduced training volume and intensity maintain aerobic capacity but not performance in distance runners. Int. J. Sports Med. 14, 33–37.

McKendry, J., Perez-Lopez, A., McLeod, M., Luo, D., Dent, J.R., Smeuninx, B., Yu, J., Taylor, A.E., Philp, A., Breen, L., 2016. Short inter-set rest blunts resistance exercise-induced increases in myofibrillar protein synthesis and intracellular signalling in young males. Exp. Physiol. 101, 866–882.

Milia, R., Roberto, S., Pinna, M., Palazzolo, G., Sanna, I., Omeri, M., Piredda, S., Migliaccio, G., Concu, A., Crisafulli, A., 2014. Physiological responses and energy expenditure during competitive fencing. Appl. Physiol. Nutr. Metab. 39, 324–328.

Millet, G.Y., Banfi, J.C., Kerherve, H., Morin, J.B., Vincent, L., Estrade, C., Geyssant, A., Feasson, L., 2011. Physiological and biological factors associated with a 24 h treadmill ultra-marathon performance. Scand. J. Med. Sci. Sports 21, 54–61.

Mourtzakis, M., Saltin, B., Graham, T., Pilegaard, H., 2006. Carbohydrate metabolism during prolonged exercise and recovery: interactions between pyruvate dehydrogenase, fatty acids, and amino acids. J. Appl. Physiol. 100, 1822–1830.

Mujika, I., Goya, A., Padilla, S., Grijalba, A., Gorostiaga, E., Ibanez, J., 2000. Physiological responses to a 6-d taper in middle-distance runners: influence of training intensity and volume. Med. Sci. Sports Exerc. 32, 511–517.

Munoz, I., Seiler, S., Alcocer, A., Carr, N., Esteve-Lanao, J., 2015. Specific intensity for peaking: is race pace the best option? Asian J. Sports Med. 6, e24900.

Myers, A.M., Beam, N.W., Fakhoury, J.D., 2017. Resistance training for children and adolescents. Transl. Pediatr. 6, 137–143.

Nakajima, T., Yasuda, T., Koide, S., Yamasoba, T., Obi, S., Toyoda, S., Sato, Y., Inoue, T., Kano, Y., 2016. Repetitive restriction of muscle blood flow enhances mTOR signaling pathways in a rat model. Heart Vessel. 31, 1685–1695.

O'Leary, T.J., Collett, J., Howells, K., Morris, M.G., 2017. Endurance capacity and neuromuscular fatigue following high- vs moderate-intensity endurance training: a randomized trial. Scand. J. Med. Sci. Sports, 27, 1648–1661.

O'Neal, E.K., Davis, B.A., Thigpen, L.K., Caufield, C.R., Horton, A.D., McIntosh, J.R., 2012. Runners greatly underestimate sweat losses before and after a 1-hr summer run. Int. J. Sport Nutr. Exerc. Metab. 22, 353–362.

Papacosta, E., Gleeson, M., Nassis, G.P., 2013. Salivary hormones, IgA, and performance during intense training and tapering in judo athletes. J. Strength Cond. Res. 27, 2569–2580.

Pope Jr., H.G., Amiaz, R., Brennan, B.P., Orr, G., Weiser, M., Kelly, J.F., Kanayama, G., Siegel, A., Hudson, J.I., Seidman, S.N., 2010. Parallel-group placebo-controlled trial of testosterone gel in men with major depressive disorder displaying an incomplete response to standard antidepressant treatment. J. Clin. Psychopharmacol. 30, 126–134.

Radak, Z., Apor, P., Pucsok, J., Berkes, I., Ogonovszky, H., Pavlik, G., Nakamoto, H., Goto, S., 2003. Marathon running alters the DNA base excision repair in human skeletal muscle. Life Sci. 72, 1627–1633.

Radak, Z., Chung, H.Y., Goto, S., 2008. Systemic adaptation to oxidative challenge induced by regular exercise. Free Radic. Biol. Med. 44, 153–159.

Ramirez-Campillo, R., Alvarez, C., Henriquez-Olguin, C., Baez, E.B., Martinez, C., Andrade, D.C., Izquierdo, M., 2014. Effects of plyometric training on endurance and explosive strength performance in competitive middle- and long-distance runners. J. Strength Cond. Res. 28, 97–104.

Ramos, J.S., Dalleck, L.C., Tjonna, A.E., Beetham, K.S., Coombes, J.S., 2015. The impact of high-intensity interval training versus moderate-intensity continuous training on vascular function: a systematic review and meta-analysis. Sports Med. 45, 679–692.

Ribeiro, P.A., Boidin, M., Juneau, M., Nigam, A., Gayda, M., 2017. High-intensity interval training in patients with coronary heart disease: prescription models and perspectives. Ann. Phys. Rehabil. Med. 60, 50–57.

Robergs, R.A., Pearson, D.R., Costill, D.L., Fink, W.J., Pascoe, D.D., Benedict, M.A., Lambert, C.P., Zachweija, J.J., 1991. Muscle glycogenolysis during differing intensities of weight-resistance exercise. J. Appl. Physiol. 70, 1700–1706.

Robertson, S.J., Joyce, D.G., 2015. Informing in-season tactical periodisation in team sport: development of a match difficulty index for Super Rugby. J. Sports Sci. 33, 99–107.

Ronnestad, B.R., Ellefsen, S., Nygaard, H., Zacharoff, E.E., Vikmoen, O., Hansen, J., Hallen, J., 2014. Effects of 12 weeks of block periodization on performance and performance indices in well-trained cyclists. Scand. J. Med. Sci. Sports 24, 327–335.

Ronnestad, B.R., Hansen, J., Thyli, V., Bakken, T.A., Sandbakk, O., 2016. 5-week block periodization increases aerobic power in elite cross-country skiers. Scand. J. Med. Sci. Sports 26, 140–146.

Rooijackers, H.M., Wiegers, E.C., van der Graaf, M., Thijssen, D.H., Kessels, R.P.C., Tack, C.J., de Galan, B.E., 2017. A single bout of high-intensity interval training reduces awareness of subsequent hypoglycemia in patients with type 1 diabetes. Diabetes 66, 1990–1998.

Rowland, T.W., Boyajian, A., 1995. Aerobic response to endurance exercise training in children. Pediatrics 96, 654–658.

Santos-Concejero, J., Billaut, F., Grobler, L., Olivan, J., Noakes, T.D., Tucker, R., 2017. Brain oxygenation declines in elite Kenyan runners during a maximal interval training session. Eur. J. Appl. Physiol. 117, 1017–1024.

Saraslanidis, P.J., Manetzis, C.G., Tsalis, G.A., Zafeiridis, A.S., Mougios, V.G., Kellis, S.E., 2009. Biochemical evaluation of running workouts used in training for the 400-m sprint. J. Strength Cond. Res. 23, 2266–2271.

Scalzo, R.L., Peltonen, G.L., Binns, S.E., Shankaran, M., Giordano, G.R., Hartley, D.A., Klochak, A.L., Lonac, M.C., Paris, H.L., Szallar, S.E., Wood, L.M., Peelor 3rd, F.F., Holmes, W.E., Hellerstein, M.K., Bell, C., Hamilton, K.L., Miller, B.F., 2014. Greater muscle protein synthesis and mitochondrial biogenesis in males compared with females during sprint interval training. FASEB J. 28, 2705–2714.

Schumann, M., Eklund, D., Taipale, R.S., Nyman, K., Kraemer, W.J., Hakkinen, A., Izquierdo, M., Hakkinen, K., 2013. Acute neuromuscular and endocrine responses and recovery to single-session combined endurance and strength loadings: "order effect" in untrained young men. J. Strength Cond. Res. 27, 421–433.

Scudese, E., Simao, R., Senna, G., Vingren, J.L., Willardson, J.M., Baffi, M., Miranda, H., 2016. Long rest interval promotes durable testosterone responses in high-intensity bench press. J. Strength Cond. Res. 30, 1275–1286.

Shankaran, M., King, C.L., Angel, T.E., Holmes, W.E., Li, K.W., Colangelo, M., Price, J.C., Turner, S.M., Bell, C., Hamilton, K.L., Miller, B.F., Hellerstein, M.K., 2016. Circulating protein synthesis rates reveal skeletal muscle proteome dynamics. J. Clin. Invest. 126, 288–302.

Shepley, B., MacDougall, J.D., Cipriano, N., Sutton, J.R., Tarnopolsky, M.A., Coates, G., 1992. Physiological effects of tapering in highly trained athletes. J. Appl. Physiol. 72, 706–711.

Skovgaard, C., Almquist, N.W., Bangsbo, J., 2018. The effect of repeated periods of speed endurance training on performance, running economy, and muscle adaptations. Scand. J. Med. Sci. Sports, 28, 381–391.

Slawinski, J., Demarle, A., Koralsztein, J.P., Billat, V., 2001. Effect of supra-lactate threshold training on the relationship between mechanical stride descriptors and aerobic energy cost in trained runners. Arch. Physiol. Biochem. 109, 110–116.

Tabata, I., Nishimura, K., Kouzaki, M., Hirai, Y., Ogita, F., Miyachi, M., Yamamoto, K., 1996. Effects of moderate-intensity endurance and high-intensity intermittent training on anaerobic capacity and VO_2max. Med. Sci. Sports Exerc. 28, 1327–1330.

Takagi, S., Kime, R., Niwayama, M., Hirayama, K., Sakamoto, S., 2017. Effects of 8 weeks' training on systemic and muscle oxygen dynamics in university rugby players. Adv. Exp. Med. Biol. 977, 43–49.

Talanian, J.L., Holloway, G.P., Snook, L.A., Heigenhauser, G.J., Bonen, A., Spriet, L.L., 2010. Exercise training increases sarcolemmal and mitochondrial fatty acid transport proteins in human skeletal muscle. Am. J. Phys. Endocrinol. Metab. 299, E180–188.

Terada, S., Tabata, I., 2004. Effects of acute bouts of running and swimming exercise on PGC-1alpha protein expression in rat epitrochlearis and soleus muscle. Am. J. Phys. Endocrinol. Metab. 286, E208–216.

Thomson, D.M., Gordon, S.E., 2005. Diminished overload-induced hypertrophy in aged fast-twitch skeletal muscle is associated with AMPK hyperphosphorylation. J. Appl. Physiol. 98, 557–564.

Tillin, N.A., Pain, M.T., Folland, J.P., 2012. Short-term training for explosive strength causes neural and mechanical adaptations. Exp. Physiol. 97, 630–641.

Tonnessen, E., Sylta, O., Haugen, T.A., Hem, E., Svendsen, I.S., Seiler, S., 2014. The road to gold: training and peaking characteristics in the year prior to a gold medal endurance performance. PLoS One 9, e101796.

Tourinho Filho, H., Pires, M., Puggina, E.F., Papoti, M., Barbieri, R., Martinelli Jr., C.E., 2017. Serum IGF-I, IGFBP-3 and ALS concentrations and physical performance in young swimmers during a training season. Growth Hormon. IGF Res. 32, 49–54.

van Loon, L.J., Greenhaff, P.L., Constantin-Teodosiu, D., Saris, W.H., Wagenmakers, A.J., 2001. The effects of increasing exercise intensity on muscle fuel utilisation in humans. J. Physiol. 536, 295–304.

Vikmoen, O., Ellefsen, S., Troen, O., Hollan, I., Hanestadhaugen, M., Raastad, T., Ronnestad, B.R., 2016. Strength training improves cycling performance, fractional utilization of VO_2max and cycling economy in female cyclists. Scand. J. Med. Sci. Sports 26, 384–396.

Vuorimaa, T., Ahotupa, M., Hakkinen, K., Vasankari, T., 2008. Different hormonal response to continuous and intermittent exercise in middle-distance and marathon runners. Scand. J. Med. Sci. Sports 18, 565–572.

Wahl, P., Mathes, S., Achtzehn, S., Bloch, W., Mester, J., 2014. Active vs. passive recovery during high-intensity training influences hormonal response. Int. J. Sports Med. 35, 583–589.

Walker, S., Hulmi, J.J., Wernbom, M., Nyman, K., Kraemer, W.J., Ahtiainen, J.P., Hakkinen, K., 2013. Variable resistance training promotes greater fatigue resistance but not hypertrophy versus constant resistance training. Eur. J. Appl. Physiol. 113, 2233–2244.

Wang, E., Solli, G.S., Nyberg, S.K., Hoff, J., Helgerud, J., 2012. Stroke volume does not plateau in female endurance athletes. Int. J. Sports Med. 33, 734–739.

Weiner, R.B., DeLuca, J.R., Wang, F., Lin, J., Wasfy, M.M., Berkstresser, B., Stohr, E., Shave, R., Lewis, G.D., Hutter Jr., A.M., Picard, M.H., Baggish, A.L., 2015. Exercise-induced left ventricular remodeling among competitive athletes: a phasic phenomenon. Circ. Cardiovasc. Imaging 8.

Widrick, J.J., Stelzer, J.E., Shoepe, T.C., Garner, D.P., 2002. Functional properties of human muscle fibers after short-term resistance exercise training. Am. J. Physiol. Regul. Integr. Comp. Physiol. 283, R408–416.

Yu, M., Blomstrand, E., Chibalin, A.V., Krook, A., Zierath, J.R., 2001. Marathon running increases ERK1/2 and p38 MAP kinase signalling to downstream targets in human skeletal muscle. J. Physiol. 536, 273–282.

CHAPTER 13

Testing

13.1 INTRODUCTION

It is extremely important to know as much as possible about our athletes and teams, and the requirements that are necessary to be successful in a given sport. Without knowing what happens with (in) the athlete's body during a 400 m race or a marathon run, we cannot create a proper training program. Without understanding the force that must be available in order to throw the hammer to 80 m, how can we design appropriate exercise sessions? Without being able to see the metabolic demands of soccer, handball, basketball, or fencing, how can we instruct players to enhance performance during training? Knowing that in freestyle wrestling the heart rate in the third period of the 2-min, fight is around 90% of the maximum and the blood lactate concentration increases up to 11–12 mmol/L (Chino et al., 2015) gives us important information to set up the proper intensity of exercise loading. Without testing, how can we be sure that the goals of different periodization units have been achieved and we don't have to make modifications to the program? It is relatively clear that we need continuous feedback from the athletes, and testing is important to evaluate the quality of coaching decisions, which should be based on the test results and the athlete's feedback. It is not easy to select the most appropriate tests, because tests must correlate well with certain components of the performance. Most of the tests which are used to measure strength, endurance, or speed components should be technically simple and easy to learn, otherwise the quality of technical execution limits the application of the test. Furthermore, it is also important to provide very similar conditions, such as time, temperature, and motivational, nutritional, and resting levels, because without these the results would be dependent on too many alternatives and not be comparable.

Moreover, monitoring the opponents' previous games builds successful tactics that lead to victory. Because of the abovementioned considerations, there is a broad spectrum of methods and equipment by which different abilities and activities can be measured during training and even during actual events. In this chapter we shall consider some important tests that assess strength, endurance, speed, flexibility, and agility. In addition, the usefulness of some biochemical biomarkers will be discussed.

13.2 TESTS FOR MEASURING STRENGTH

There is a vast number of tests by which maximal strength, explosive strength, or endurance strength can be assessed. Some of these tests only require body weight and free loads, while others require the use of sophisticated machines.

The hand grip test is mainly used to assess general strength and health status of populations, in epidemiological studies. In large population studies, hand grip results are associated with mortality in aging populations and/or in patients with various diseases (Bohannon, 2008). However, this test is not very useful for elite athletes due to the poor correlation with sport performance.

One of the simplest and most often used tests in strength training is the determination of one maximum repetition (1 MR) for a given exercise—a measure of the maximal weight a subject can lift only once for any given lift. Let us take, for example, the bench press, which is an important test for a shot-putter and correlates highly with sport performance. From the number of repetitions at 80% of 1 MR, we have a clue as to the fiber type composition of the triceps brachii muscle. If one can do fewer than seven repeats at 80% it is probably due to fast fiber dominance, due to the rapid fatigue. On the other hand, if one is capable of 12 or more repetitions, this would be a reflection of probably slow-twitch fiber dominance, because of the late onset of fatigue. Regular measurements of 1 MR in exercises that relate to sport, such as squats, which are associated with the performance of clean and jerk and hammer throw, or bench pull which is linked to the performance of kayakers, are important because it is an easy assessment of maximal strength. The number of repetitions at given weight, such as 70% or 80%, is often used to evaluate training adaptation.

Explosive strength is easily evaluated by measuring vertical or horizontal jumps from a standing position or from initial movements. The objectivity of this test can be increased by measuring vertical and horizontal reaction forces on a force platform. These platforms make it possible to test the generated force during real sport movements, like long jump and high jump, track and field throwing events, team-handball jump throw, volleyball jump for block and spike, and judo throws, among others. From the results one can precisely measure the generated force, dynamics, mechanical energy, etc. Drop jump is often used in training sessions aiming to increase explosive strength. One can start jumping down and then up from 20 cm, and increasing the height of the bench until the performance (height) of the drop jumps increase. Very well-trained athletes can increase the maximal height of the drop jump from a bench as high as 100 cm. The exact height, force, and power can be measured or calculated if the jumps are done on a force plate.

Explosive strength is readily measured by the distance a medicine ball or kettlebell can be thrown. However, different types of accelerometers can be used to evaluate the degree of explosive strength. Moreover, accelerometers are often used to pinpoint the load in which the maximal power is achieved. For instance, pull-ups done by resistance exercise-trained male subjects maximized power at 71% of 1 MR (Munoz-Lopez et al., 2017). In another study, the maximal power output for bench press by soccer players attained the maximal value at around 50% and the data showed significant individual differences (Jandacka and Uchytil, 2011). Therefore, exact measurements of maximal power output are important for individual load setting, but can be reasonably difficult to compare between athletes from different sports.

The measurements of vertical jump performance before and immediately after various ball games, like soccer, team-handball, volleyball, or basketball, would give important information about how explosive strength changes as a result of the game. Significant decreases in the jump height by a given player would provide clues regarding the need for individualized training under these conditions.

Strength endurance can be readily assessed by calculating the number of repetitions at different intensities, but is generally 40%–60% of 1 MR. Strength endurance is often measured by field tests also. For example, the generated force curve on the blade of a kayak paddle at different distances in a 1000 m race could be very useful information for the coach when re-assessing training sessions.

In rowing, for example, force, power, and velocity of the oar can be readily measured using an appropriate ergometer in the laboratory. However, the characteristic of the force curve assessed on the blade during actual rowing provides information about the biomechanics of the given technique. The laboratory assessment is important because of the reproducibility of conditions (same machine, temperature, humidity, etc.) while doing measurements on water, despite the constantly changing conditions like wind, stream, temperature, etc., are more valuable since they show competition data. Therefore, both laboratory and field tests are important to evaluate the progress of preparation, the level of physical fitness, and the success of regeneration.

13.3 TESTS TO ASSESS ENDURANCE

Cardiorespiratory fitness is measured by VO_2max which is one of the most often used tests to evaluate endurance capacity. VO_2max is generally measured in laboratories using treadmill running, cycling, or rowing ergometers by progressively increasing intensity over a time period that exceeds 5 min. Heart rate, ventilation, and inhaled and exhaled oxygen and carbon dioxide differences are measured. The cardiac output and the arteriovenous oxygen difference multiplication factor gives the value of VO_2max which, divided by body weight (mL/kg/min), which can be used to compare athlete results from the same or different sports and even the progress made over training sessions by an individual athlete. With the help of a portable mask, VO_2max can be measured outdoors and in the competition setting, during running, skiing, cycling, kayaking, rowing, and even swimming.

Because of the importance of VO_2max in health, since the value is negatively correlated with the incidence of a wide range of diseases, and in elite sport, since certain values are almost obligatory for success in endurance-associated sports, many methods have been developed to estimate VO_2max.

The Cooper running test, 12 min of continuous running, is a good estimate of VO_2max. Cooper's equation, VO_2max = running distance−504.9/44.73, is now well-established and used around the world.

In well-trained subjects, aged 21–51 years, the VO_2max can be judged efficiently by the differences in the maximal and resting heart rates multiplied by 15.3 (Uth et al., 2004): VO_2max = HRmax/HRrest × 15.3.

Another popular estimation of VO_2max is the Rockport Fitness Walking Test, which uses 1.6 km (1 mile, 4 laps of a standard track) with the subjects walking as fast as possible, to estimate the VO_2max using the following equation: VO_2max = 132.853 − (0.0769 × body weight in pounds) − (0.3877 × age) + (6.315 × 0 if you are female or 1 if you are male) − (3.2649 × time to complete 1.600 m in minutes) − (0.1565 × number of heart beats in 10 s at the end of the 1-mile walk).

The bench step test, using a 42 cm-high bench, a stopwatch, and a metronome is another relatively straightforward test to estimate VO_2max. The subject steps up and down, 1 ft at a time, onto the bench for 3 min, at a rate of 22 steps/min. For females and 24 for males. The number of heartbeats in the last minute is recorded using a heart rate monitor. Then, use the equation: $VO_2max = 111.33 - (0.42 \times last\ minute\ heart\ rate)$ for males and $VO_2max = 65.81 - (0.1847 \times last\ minute\ heart\ rate)$ for females, to estimate VO_2max.

VO_2max is important for cardiovascular and muscular endurance. However, the economy of the exercise is also very important. With the same VO_2max, a competitor who can work at a higher percentage of VO_2max in aerobic conditions has an advantage. Within a certain range, athletes who are able to work at a higher percentage of VO_2max can easily compensate for a lower level of VO_2max. Therefore, without question, one of the main goals of endurance training is to increase the intensity at which athletes reach the anaerobic threshold, which can be measured in the laboratory. The measurement allows one to determine heart rate and the intensity of the movement (running, cycling, rowing) at the point where the lactate levels increase in a steeper pattern, generally around 4 mmol/L (see Chapter 5).

The determination of lactate threshold can be done in the laboratory, using a treadmill or cycle ergometer, but a sport-specific field test could be more valuable. To measure lactate threshold, at least four intensities—low, moderate, high, and maximal intensity—must be selected in which heart rate and lactate levels are measured (Fig. 13.1). As an example, swimmers are asked to swim 200 m with low intensity, then their heart rate and blood lactate would be measured. This procedure would be repeated after moderate, high, and maximal intensities of swimming. The obtained results would allow estimation of the heart rate at the blood lactate level of 4 mmol/L. We can then use this value to set up the intensity based on the heart rate of a swimmer.

The Beep Test is widely used to assess endurance especially in ball games. In this test, participants must run between two lines 20 m apart and reach the lines before recorded signals beep. The test starts with a slow speed, which is gradually increased as the periods between beeps get shorter and shorter. If the line is reached before the beep sounds, the subject must wait until the beep sounds before continuing; and if participants do not reach the line (within 2 m) for two consecutive runs there is a warning, and on the third failure to reach the line in time, the test ends.

The Yo-Yo Test is similar to the Beep Test. The intermittent recovery 1 (IR-1) is created for recreational players starts at a speed of 10 km/h, while version 2 (IR-2) starts at 13 km/h and is designed for elite players. The Yo-Yo Test consists of 2×20 m shuttle runs at increasing speeds, interspersed with a 10-s period of active recovery (controlled by audio signals from a

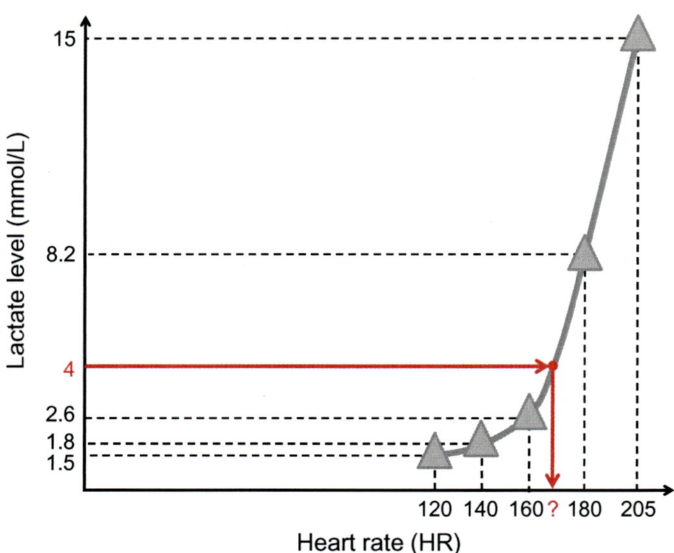

Fig. 13.1 Schematic model to measure lactate threshold. The measurement of lactate threshold-associated intensity and heart rate is crucial to the proper design of endurance training. Field tests are preferred to laboratory tests, and this figure shows a protocol in which four exercise intensity zones (low, moderate, high, and maximal) are used with measurements of heart rate and lactate levels. The connection of lactate and heart rate values after the completion of exercise with four intensity zones provide the heart rate value which belongs to 4 mmol/L lactate levels. The intensity below this heart rate would represent a zone in which the ATP production is mostly produced by aerobic metabolism, and above this heart rate the energy is produced at greater degree by anaerobic metabolism. This test has to be repeated regularly and the heart rate has to be adjusted according to the progress of adaptation.

compact disc). The test is terminated when an individual is no longer able to maintain the required speed. The distance covered up to the end-point is represented in the test results (Iaia et al., 2008). VO$_2$max can be judged by the distance of the Yo-Yo Test using the following formula:

$$Yo-Yo-IR-1: VO_2\,max = distance\ in\ m \times 0.0084 + 36.4.$$

$$Yo-Yo-IR-2: VO_2\,max = distance\ in\ m \times 0.0136 + 45.3.$$

However, Yo-Yo-IR tests have been shown to be a more sensitive measure of change in performance than has maximum oxygen uptake (Bangsbo et al., 2008).

The game-based performance test (GBPT) for team-handball was developed by Wagner et al. (2016) and consists of incremental treadmill running and a team-handball test game (TG) (2×20 min) peak oxygen uptake ($\dot{V}O_2$ max), blood lactate concentration (BLC), heart rate (HR), sprinting time, time of offensive and defensive actions, as well as running intensities, ball velocity, and jump height tests. This test aims to evaluate the most of the important elements of team-handball. However, in order to be able to see the validity of this test, it must be applied widely and a correlation seen with real performance. In ball games and combat sports, coaches use objective (measured) and subjective scales (skillfulness, game-IQ, etc.) to evaluate players. Both are important.

13.4 TESTS TO EVALUATE SPEED

Fiber type ratio, number, and hypertrophy are powerful determinants in maximal speed-related sports. Fiber type ratio is dependent on genetics, and the measurements of fiber type ratio show differences in the skeletal muscle of endurance and explosive sport-involved athletes. The dominance of fast-twitch fibers is clearly evident in explosive sports, while endurance athletes enjoy the benefits of higher slow-twitch fibers. Gollnick and his colleagues in the early 1970s showed training associated differences in fiber types (Gollnick et al., 1972, 1973). The gold standard for the measurements of fiber types is the muscle biopsy, but noninvasive methods have been used to evaluate fiber types. NMR generates nuclear magnetization and the changes in this signal, called relaxation, are characterized by T1 and T2. T1 is responsible for loss of signal density while T2 is a measure of a broadening of the signals. Higher proportions of fast-twitch fibers are associated with longer relaxation times, which suggest that NMR can be used to assess fiber types (Kuno et al., 1988). Another noninvasive method, the proton magnetic resonance spectroscopy (^{1}H-MRS) has been used to measure fiber types, based on the significant differences in carnosine content in slow- and fast-twitch fibers (Baguet et al., 2011). A significant positive correlation was found between muscle carnosine levels and percentage area occupied by type II fibers. Explosive athletes had ~30% higher carnosine levels compared to a reference population, whereas it was ~20% lower than normal in typical endurance athletes (Baguet et al., 2011).

Fractional anisotropy can be measured by diffusion tensor imaging, and it has been shown that higher fractional anisotropy correlated well with slow-twitch fiber ratio (Scheel et al., 2013), adding another tool to noninvasive appraisal of fiber types.

Not just imaging methods, but simple measurements of the delay, contraction, and half relaxation times of the muscle can be used to investigate fiber types, because these parameters correlate with the proportion of myosin heavy chains (Simunic et al., 2011). Skeletal muscle mechanical radial twitch response can be applied as an accurate noninvasive predictor of fiber types (Simunic et al., 2011). Assessment of fibers types could be important for the selection of an appropriate sport in the early phase of athletic development. However, it must be noted that fiber types can be altered, within a range, by training (Ellefsen et al., 2014; Raue et al., 2012), and within a few sports, elite athletes can be developed regardless of fiber type distribution.

The Wingate Test was created at the Wingate Institute and the aim was to measure anaerobic power and anaerobic capacity, usually using a cycle ergometer. After proper warming up, and a rest before the start, subjects pedal as fast as they can for 30 s (Dotan and Bar-Or, 1980). Ergometers collect the date and display, through a computer, which calculate peak anaerobic power, anaerobic capacity, and anaerobic fatigue. Results from the Wingate Test correlated with the performance test of 50 and 400 m free style swimming (Hawley et al., 1992). Wingate Test results are good predictors of performance for 1500 m speed skating (Hofman et al., 2017) and wrestling (Garcia-Pallares et al., 2011), but no relationship was found between Wingate Test results and water polo performance (Bampouras and Marrin, 2009) running performance (100–3000 m) (Legaz-Arrese et al., 2011), or table tennis (Zagatto et al., 2008), which demonstrates the complexity of sport performance.

The Specific Aerobic Gymnast Anaerobic Test was created by Alves et al. (2015) to asses anaerobic capacity of gymnasts in a sport-specific manner.

To achieve greater speed of release in throwing events of track and field is one of the main aims of athletes, because of its relation to throwing distance. Therefore, using lighter weight than the original competitive ones is an effective way to improve speed in many sports. Using lighter objects in shotput (6 kg instead of 7.25 kg), discuss throwing (1.75 kg instead of 2 kg), or javelin throw (700 g instead of 800 g) could be a good tool to improve speed and/or test sport-specific speed of athletes. Heavier objects are also used to improve sport-specific explosive strength.

13.5 FLEXIBILITY ASSESSMENT

Range of motion limits performance in many sports, and this can be closely related to injury prevention as well as demonstrate the progress of recovery from injury. A large number of flexibility tests are available, but only a few will be mentioned in this section. Readers are referred to a number of excellent reviews on this topic (Dallinga et al., 2012; Ellenbecker and Cools, 2010; Erickson et al., 2015; Essendrop et al., 2002; Martin et al., 2010; Mayorga-Vega et al., 2014; Reiman et al., 2015).

Goniometers and digital protractors are designed to measure range of motion. Proper positioning of each goniometer aids objective measurement of range of motion (ROM) at a given joint. However, most flexibility tests measure the reaching point from different positions for different joints. The maximal reaching point, in the sitting position and leaning forward with stretched knees, tests the flexibility of the *biceps femoris* muscle (hamstring) and lower back muscles. The same muscle flexibility can be measured using a standing test, where from a standing position, the athlete leans down to touch the floor with their hands without bending the knees. Hamstring injuries are common in short-distance running and ball games, and can often be long term and debilitating. The active knee extension test is a reliable and valid assessment of posterior thigh flexibility; during the test, athletes are positioned supine in the 90/90 position on an examination couch, with both knee and hip flexed to 90°. The opposite leg is placed flat on the couch with the knee fully extended and maintained in this position throughout the test. The athlete is asked to actively extend the knee being tested through the full available ROM until firm muscular resistance is felt, while the hip is maintained at 90° flexion. The stationary arm of an inclinometer (which measures angles) or goniometer is aligned along the femur. Angles, in degrees, are measured on both sides, representing the ROM. This test is often used to evaluate the active ROM in the knee and provides valuable information on the risk of hamstring injury (Malliaropoulos et al., 2015).

The glenohumeral joint is under a great load in javelin throwers, baseball pitchers, and handball and tennis players. These athletes demonstrate a loss of internal rotation and a gain in external rotation in the throwing or serving arm. To test the level of internal rotation, the athlete is asked to lie supine on the plinth. The physiotherapist positions the shoulder at 90° abduction and 90° elbow flexion with the arm in neutral rotation. The tester moves the arm of the athlete into internal rotation until a capsular end feel is achieved, or the subject complains of pain. After recording the value, the arm is returned to neutral rotation. For external rotation the tester resets the arm into the starting position of 90° of abduction and 90° of elbow flexion with neutral rotation. The arm is then moved into maximal external rotation, with the plinth providing scapular stabilization, until the subject obtains a capsular end feel or the participant complains of pain (Kevern et al., 2014).

The results of this test show the degree of range of motion during internal and external rotation of the glenohumeral joint.

The seated rotation test is used to evaluate the range of trunk rotation. Athletes are asked to sit in a chair with feet positioned together on the ground, the body in an erect upright posture with arms across the chest. Subjects are then instructed to rotate to the right as far as possible without discomfort and, with the help of a goniometer, the degree of rotation is measured and recorded. Then the test is repeated with rotation to the left side. It is advisable to repeat all flexibility tests three times, using the average value to increase the reliability of the testing.

A small degree of range of motion and hypermobility can both be risk factors for injury. Indeed, generalized joint laxity is an intrinsic risk factor for noncontact anterior cruciate ligament injuries (Alentorn-Geli et al., 2009), and hypermobility often displays a lax patellofemoral joint with or without a history of subluxation or dislocation. The Beighton and Horan Joint Mobility Index is commonly used to assess generalized joint laxity, and its validity and reliability have been established. This widely used test consists of five objective measurements of joint mobility, four of which are measured bilaterally: elbow extension, knee extension, little finger extension and thumb extension, and the forward bending reaching test. The detection of hypermobility in these tests gives the mobility index (Jindal et al., 2016).

13.6 AGILITY TESTS

Agility is the capacity to change course, controlling the direction and position of one's body while maintaining momentum. It is important in all ball games and combat sports. There are numerous methods to test agility and many of them were developed for specific sports. Using the most accepted test for a given sport offers the possibility to compare test results with other players. Indeed, tests are not just used to measure individual progress but, if the given tests are used widely, the results can be compared in larger sport-specific populations. Below are short descriptions of some agility tests that are used internationally.

The Balsom Agility Test is designed for soccer players, and mimics the changes in running direction during a soccer game. The course of this test is shown in Fig. 13.2. Players start at point A, and sprint to the cones at point B. They turn at point B, sprint back through point A, turn to the left, and sprint through point C to point D. They turn at point D and then sprint back through C, turn to the right, and sprint through point B to the

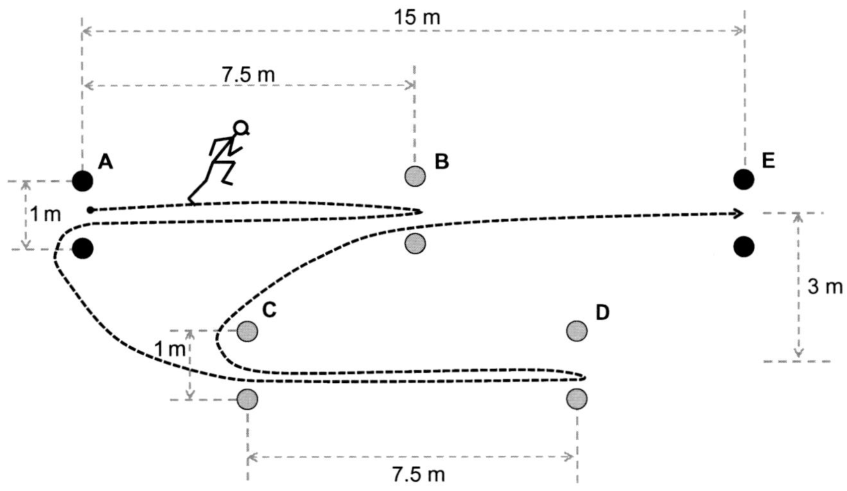

Fig. 13.2 The Balsom Agility Test course.

finishing gate shown at point E (Fig. 13.2). This test is widely used, which provides the possibility of comparing the results of players at different levels.

The Arrowhead Agility Test got its name from the shape of placed markers that the players have to cover, i.e., three marker cones placed in an arrowhead shape, and one set of cones or line marker to indicate the start and finish lines. The players run as fast as possible from the start line to the middle cone (A), turn to run around the cone (C) or (D), around the far cone (B), and back through the start/finish line. The best time out of four trials is registered. This test is also often used in soccer and team-handball, and is helpful to evaluate the objective agility of the players (Flotum et al., 2016).

The Lateral Change Direction Test aims to test the agility required for basketball and handball players. Three cones (A, B, C) are placed in a line, 5 m apart. The player stands behind the middle cone (B) and on a sign from an assistant moves either right (C) or left (A). The player touches the first designated cone, then returns past the middle cone to the far cone and touches it, and then returns to and touches the middle cone to complete the test. The shortest time out of four trials is registered.

The Star Agility Test has been shown to be appropriate for children aged 8–10 (Golle et al., 2015). They are asked to run using different running techniques (e.g., forward, backward, side steps) from the center of a 9×9-m star-shaped field to the edge and back with four markers. The time of this trial is measured. This agility test is often used in long-term development in various sports, including ball games.

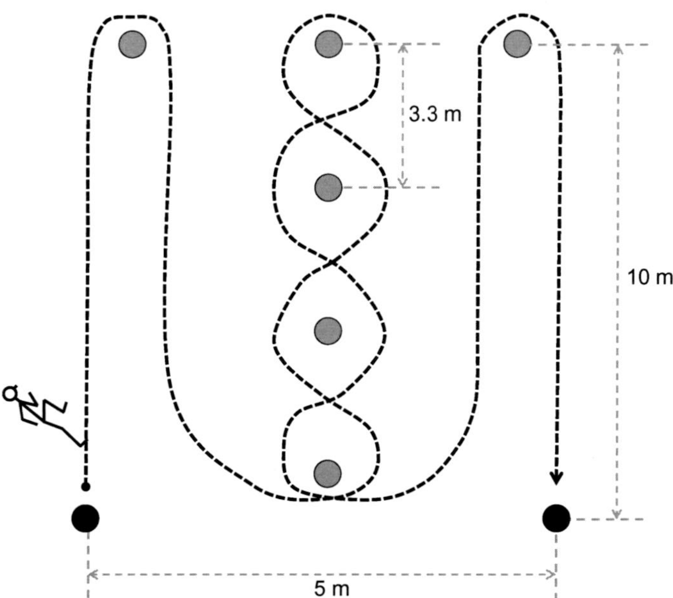

Fig. 13.3 The Illinois Agility Test course.

The Illinois Agility Test measures the ability to turn in different directions and at different angles in a course which is marked by cones (10 m long and 5 m wide) (Fig. 13.3). Each subject starts face down, in a prone position, with the head at the start line and hands by the shoulders; they wait for the signal of an assistant to start to run over the course, without knocking down any cones.

The *t*-test measures the ability to change direction at high speed. Four cones are arranged in a T-shape, with cone B placed 9.14 m from cone A, and two additional cones (C and D) placed 4.57 m on either side of cone B. Players starts to run, on a sound signal, forward from A to B, shuffle 4.57 m to the left to cone C, shuffle 9.14 m to the right to cone D, and shuffle 4.57 m back left to cone B before finally running backwards to reach cone A. The time is recorded.

13.7 BIOCHEMICAL BIOMARKERS

Biomarkers used for personalized medicine are typically categorized as either prognostic or predictive. It is not easy to define the meaning of biomarker, but in 1998, the National Institutes of Health Biomarkers Definitions Working Group defined a biomarker as "a characteristic that is objectively measured and evaluated as an indicator of normal biological processes, pathogenic processes, or pharmacologic responses to a therapeutic intervention" (Lozano de Luaces et al., 1988). Biomarkers in exercise and sport are used for diagnosis or prediction and to assess the degree of adaptation. However, because of the complexity of performance in exercise, the correlations between the biochemical biomarkers and performance are not always very strong. As a result, the applicability of certain biomarkers is under debate.

Biomarkers of anabolic metabolism: Despite the fact that lactic acid is produced even during sleep, there is a strong relationship between the intensity of exercise and lactic acid concentration in the blood. Similar to the relationship with lactic acid, the concentration of ammonia increases with the intensity of exercise and is also used as a marker of anaerobic metabolism.

Muscle damage: Creatine kinase (CK) and lactate dehydrogenase (LDH) are enzymes of relatively small size. Hence, they can easily efflux from the working muscle into the circulation and are often used as markers of muscle damage. When subjects with elevated levels of CK and subjects with normal levels of CK were subjected to incremental exercise, data revealed similar aerobic capacities and cardiovascular responses. On the other hand, subjects with $CK \geq 300\,U/L$ exhibited significantly higher lactate and ammonia levels after maximal exercise, together with increased ventilatory responses, whereas those with $CK \geq 200\,U/L$ but $\leq 300\,U/L$ did not (Cooper et al., 2017). Prolonged exercise-associated increases in CK concentration are more closely related to the distance of the race than the level of physical fitness (Shin et al., 2016). Muscle damage is also assessed by the circulating levels of myoglobin (Niemela et al., 2016). A short-term increase of these proteins in the circulation is more or less normal, and could be due to the increased permeability of sarcolemma. On the other hand, chronic increases could be a signal of prolonged muscle damage. This is also true for the increased concentration of uric acid, which is a purine metabolite, and therefore can be increased after high intensity exercise. There are reports that suggest that during overtraining uric acid concentration is chronically elevated (Fatouros et al., 2006).

Overtraining: As mentioned earlier in this book, currently we do not have specific and reliable biomarkers for overtraining. Because during overtraining the protein breakdown is elevated compared to the level of protein synthesis, urea levels could provide useful information. Since urea is mostly formed in the liver as a waste product from the breakdown of proteins, exhaustive training and overtraining can lead to elevated urea levels in the circulation (the normal value is around 6 mmol/L). The ratio of testosterone and cortisol hormones is also considered to be a marker of overtraining.

Cardiac biomarkers: Cardiac troponin T (cTnT) is a highly sensitive and specific indicator of cardiomyocyte necrosis, even in the presence of skeletal myocyte damage. Several studies have shown significant increases in plasma levels of cTnT or cardiac troponin I in the early postrace period following endurance events (Wanaka et al., 1989). In addition to troponin T, the N-terminal of the prohormone brain natriuretic peptide is also used to evaluate cardiac damage after exercise (Block, 1970). Homocysteine, which is an amino acid derived from methionine, is considered to be a marker of a cardiac load caused by exercise (Stewart et al., 2017).

13.7.1 Biomarkers of Liver and Kidney Function During Exercise Training

Physical exercise, when strenuous or prolonged, has been shown to increase circulating alanine aminotransferase and aspartate aminotransferase levels, indicating a degree of liver trauma. Gamma-glutamyl-transferase, glutamate dehydrogenase, and γ-GTP also serve as specific markers for liver injury, and their levels are increased after long-distance running (Nagel et al., 1990). The increase in bilirubin after long-distance running is caused by hemocytocatheresis, physiological turnover, and catabolism of hemoglobin, or by liver injury. Therefore, liver function is often tested in relation to exercise training (De Paz et al., 1995).

Creatinine is the end-product of muscle metabolism derived from creatine degradation. The clearance of creatinine is achieved by the kidneys through glomerular filtration. Blood urea nitrogen (BUN) is also used to evaluate exercise-associated alteration of kidney function. The increases in BUN during severe exercise, of long duration, may be a result of the prolonged accumulation of nonprotein nitrogen (Shin et al., 2016).

13.8 "R" EXTRA

Decision-making speed is important in all combat sports and ball games. There is a variety of methods to measure this ability, and some of them mimic field situations. In soccer, with the help of a large screen and a video recording system with game situation field-specific decision-making time, speed can be assessed. The participants are required to move to a previously defined mark 5 m in front of a large screen (240 cm × 170 cm) and to do so within 3 s of the termination of the respective exercise intensity mode. A ball is placed 50 cm from the mark and the central axis of the body. Participants start this test from a standing position. A 3 s video of an offensive game situation is played. After the video, the players select the most accurate motor response as quickly as possible, and perform the motor response in the shortest time possible. Visual-motor response time is recorded. The motor response comprises the following seven variants of offensive game activities: forward pass, pass to the right, pass to the left, diagonal pass to the right, diagonal pass to the left, beating a defender, and dribbling with the ball. The accuracy of the motor response and visual-motor response time is recorded and analyzed from the video recording, which simultaneously captures the moment when the video of the offensive game situation stopped and the moment when a lower limb touched the ball. Visual-motor response time is referred to as a period of time from the moment when the video of the offensive game situation stopped to the moment when the ball touched the participant's lower limb. The accuracy of the motor response is assessed by the choice of one of the seven motor response alternatives above.

13.9 SUMMARY

Regular testing is very important in sport because it helps to assess the progress of training, analyzes the effects of exercise on the body, and allows comparison of the tests results with others. Tests must correlate with the ability that we hope to measure. Tests must be simple, valid, and readily reproducible. Due to the extreme complexity of sport performance, it is difficult to find tests that completely reflect or predict the performance in complex sports like soccer, tennis, basketball, fencing, etc. Therefore, tests are designed to measure abilities that are important to be successful in these sports. It is important to construct sport-specific field tests, which can reflect the degree of abilities that are important for a given sport.

TEST QUESTIONS

1. How does one select tests to evaluate specific abilities for a given sport?
2. Name three tests that measure aerobic capacity.
3. Name three tests that measure anaerobic capacity.
4. How can you measure speed in ball games?
5. How can you evaluate range of motion?
6. What tests assess agility?
7. What are the limitations of biochemical biomarkers?
8. Describe some biochemical markers that are relevant to specific sports and explain why.

BIBLIOGRAPHY

Alentorn-Geli, E., Myer, G.D., Silvers, H.J., Samitier, G., Romero, D., Lazaro-Haro, C., Cugat, R., 2009. Prevention of non-contact anterior cruciate ligament injuries in soccer players. Part 1: Mechanisms of injury and underlying risk factors. Knee Surg. Sports Traumatol. Arthrosc. 17, 705–729.

Alves, C.R., Borelli, M.T., Paineli Vde, S., Azevedo Rde, A., Borelli, C.C., Lancha Junior, A.H., Gualano, B., Artioli, G.G., 2015. Development of a specific anaerobic field test for aerobic gymnastics. PLoS One 10, e0123115.

Baguet, A., Everaert, I., Hespel, P., Petrovic, M., Achten, E., Derave, W., 2011. A new method for non-invasive estimation of human muscle fiber type composition. PLoS One 6, e21956.

Bampouras, T.M., Marrin, K., 2009. Comparison of two anaerobic water polo-specific tests with the Wingate test. J. Strength Cond. Res. 23, 336–340.

Bangsbo, J., Iaia, F.M., Krustrup, P., 2008. The Yo-Yo intermittent recovery test: a useful tool for evaluation of physical performance in intermittent sports. Sports Med. 38, 37–51.

Block, G.E., 1970. Successful extended hemipelvectomy for "inoperable" chondrosarcoma of the pelvis. Surg. Clin. North Am. 50, 985–997.

Bohannon, R.W., 2008. Hand-grip dynamometry predicts future outcomes in aging adults. J. Geriatr. Phys. Ther. 31, 3–10.

Chino, K., Saito, Y., Matsumoto, S., Ikeda, T., Yanagawa, Y., 2015. Investigation of exercise intensity during a freestyle wrestling match. J. Sports Med. Phys. Fitness 55, 290–296.

Cooper, C.B., Dolezal, B.A., Neufeld, E.V., Shieh, P., Jenner, J.R., Riley, M., 2017. Exercise responses in patients with chronically high creatine kinase levels. Muscle Nerve 56, 264–270.

Dallinga, J.M., Benjaminse, A., Lemmink, K.A., 2012. Which screening tools can predict injury to the lower extremities in team sports?: a systematic review. Sports Med. 42, 791–815.

De Paz, J.A., Villa, J.G., Lopez, P., Gonzalez-Gallego, J., 1995. Effects of long-distance running on serum bilirubin. Med. Sci. Sports Exerc. 27, 1590–1594.

Dotan, R., Bar-Or, O., 1980. Climatic heat stress and performance in the Wingate anaerobic test. Eur. J. Appl. Physiol. Occup. Physiol. 44, 237–243.

Ellefsen, S., Vikmoen, O., Zacharoff, E., Rauk, I., Slettalokken, G., Hammarstrom, D., Strand, T.A., Whist, J.E., Hanestadhaugen, M., Vegge, G., Fagernes, C.E., Nygaard, H., Hollan, I., Ronnestad, B.R., 2014. Reliable determination of training-induced alterations in muscle fiber composition in human skeletal muscle using quantitative polymerase chain reaction. Scand. J. Med. Sci. Sports 24, e332–342.

Ellenbecker, T.S., Cools, A., 2010. Rehabilitation of shoulder impingement syndrome and rotator cuff injuries: an evidence-based review. Br. J. Sports Med. 44, 319–327.

Erickson, B.J., Frank, R.M., Harris, J.D., Mall, N., Romeo, A.A., 2015. The influence of humeral head inclination in reverse total shoulder arthroplasty: a systematic review. J. Shoulder Elb. Surg. 24, 988–993.

Essendrop, M., Maul, I., Laubli, T., Riihimaki, H., Schibye, B., 2002. Measures of low back function: a review of reproducibility studies. Clin. Biomech. (Bristol, Avon) 17, 235–249.

Fatouros, I.G., Destouni, A., Margonis, K., Jamurtas, A.Z., Vrettou, C., Kouretas, D., Mastorakos, G., Mitrakou, A., Taxildaris, K., Kanavakis, E., Papassotiriou, I., 2006. Cell-free plasma DNA as a novel marker of aseptic inflammation severity related to exercise overtraining. Clin. Chem. 52, 1820–1824.

Flotum, L.A., Ottesen, L.S., Krustrup, P., Mohr, M., 2016. Evaluating a nationwide recreational football intervention: recruitment, attendance, adherence, exercise intensity, and health effects. Biomed. Res. Int. 2016, 7231545.

Garcia-Pallares, J., Lopez-Gullon, J.M., Muriel, X., Diaz, A., Izquierdo, M., 2011. Physical fitness factors to predict male Olympic wrestling performance. Eur. J. Appl. Physiol. 111, 1747–1758.

Golle, K., Muehlbauer, T., Wick, D., Granacher, U., 2015. Physical fitness percentiles of German children aged 9–12 years: findings from a longitudinal study. PLoS One 10, e0142393.

Gollnick, P.D., Armstrong, R.B., Saubert, C.W., Piehl, K., Saltin, B., 1972. Enzyme activity and fiber composition in skeletal muscle of untrained and trained men. J. Appl. Physiol. 33, 312–319.

Gollnick, P.D., Armstrong, R.B., Saltin, B., Saubert, C.W., Sembrowich, W.L., Shepherd, R.E., 1973. Effect of training on enzyme activity and fiber composition of human skeletal muscle. J. Appl. Physiol. 34, 107–111.

Hawley, J.A., Williams, M.M., Vickovic, M.M., Handcock, P.J., 1992. Muscle power predicts freestyle swimming performance. Br. J. Sports Med. 26, 151–155.

Hofman, N., Orie, J., Hoozemans, M.J., Foster, C., de Koning, J.J., 2017. Wingate test is a strong predictor of 1500 m performance in elite speed skaters. Int. J. Sports Physiol. Perform, 1–17.

Iaia, F.M., Thomassen, M., Kolding, H., Gunnarsson, T., Wendell, J., Rostgaard, T., Nordsborg, N., Krustrup, P., Nybo, L., Hellsten, Y., Bangsbo, J., 2008. Reduced volume but increased training intensity elevates muscle Na+-K+ pump alpha1-subunit and NHE1 expression as well as short-term work capacity in humans. Am. J. Physiol. Regul. Integr. Comp. Physiol. 294, R966–974.

Jandacka, D., Uchytil, J., 2011. Optimal load maximizes the mean mechanical power output during upper extremity exercise in highly trained soccer players. J. Strength Cond. Res. 25, 2764–2772.

Jindal, P., Narayan, A., Ganesan, S., MacDermid, J.C., 2016. Muscle strength differences in healthy young adults with and without generalized joint hypermobility: a cross-sectional study. BMC Sports Sci. Med. Rehabil. 8, 12.

Kevern, M.A., Beecher, M., Rao, S., 2014. Reliability of measurement of glenohumeral internal rotation, external rotation, and total arc of motion in 3 test positions. J. Athl. Train. 49, 640–646.

Kuno, S., Katsuta, S., Inouye, T., Anno, I., Matsumoto, K., Akisada, M., 1988. Relationship between MR relaxation time and muscle fiber composition. Radiology 169, 567–568.

Legaz-Arrese, A., Munguia-Izquierdo, D., Carranza-Garcia, L.E., Torres-Davila, C.G., 2011. Validity of the Wingate anaerobic test for the evaluation of elite runners. J. Strength Cond. Res. 25, 819–824.

Lozano de Luaces, V., Espias Gomez, A., Murtra Ferre, J., Ruano Gil, D., 1988. Alopecia areata: review and presentation of some clinical cases. Av. Odontoestomatol. 4, 273–279.

Malliaropoulos, N., Kakoura, L., Tsitas, K., Christodoulou, D., Siozos, A., Malliaras, P., Maffulli, N., 2015. Active knee range of motion assessment in elite track and field athletes: normative values. Muscles Ligaments Tendons J. 5, 203–207.

Martin, H.D., Shears, S.A., Palmer, I.J., 2010. Evaluation of the hip. Sports Med. Arthrosc. 18, 63–75.

Mayorga-Vega, D., Merino-Marban, R., Viciana, J., 2014. Criterion-related validity of sit-and-reach tests for estimating hamstring and lumbar extensibility: a meta-analysis. J. Sports Sci. Med. 13, 1–14.

Munoz-Lopez, M., Marchante, D., Cano-Ruiz, M.A., Chicharro, J.L., Balsalobre-Fernandez, C., 2017. Load, force and power-velocity relationships in the prone pull-up exercise. Int. J. Sports Physiol. Perform, 1–22.

Nagel, D., Seiler, D., Franz, H., Jung, K., 1990. Ultra-long-distance running and the liver. Int. J. Sports Med. 11, 441–445.

Niemela, M., Kangastupa, P., Niemela, O., Bloigu, R., Juvonen, T., 2016. Individual responses in biomarkers of health after marathon and half-marathon running: is age a factor in troponin changes? Scand. J. Clin. Lab. Invest. 76, 575–580.

Raue, U., Trappe, T.A., Estrem, S.T., Qian, H.R., Helvering, L.M., Smith, R.C., Trappe, S., 2012. Transcriptome signature of resistance exercise adaptations: mixed muscle and fiber type specific profiles in young and old adults. J. Appl. Physiol. 112, 1625–1636.

Reiman, M.P., Mather 3rd, R.C., Cook, C.E., 2015. Physical examination tests for hip dysfunction and injury. Br. J. Sports Med. 49, 357–361.

Scheel, M., von Roth, P., Winkler, T., Arampatzis, A., Prokscha, T., Hamm, B., Diederichs, G., 2013. Fiber type characterization in skeletal muscle by diffusion tensor imaging. NMR Biomed. 26, 1220–1224.

Shin, K.A., Park, K.D., Ahn, J., Park, Y., Kim, Y.J., 2016. Comparison of changes in biochemical markers for skeletal muscles, hepatic metabolism, and renal function after three types of long-distance running: observational study. Medicine (Baltimore) 95, e3657.

Simunic, B., Degens, H., Rittweger, J., Narici, M., Mekjavic, I.B., Pisot, R., 2011. Noninvasive estimation of myosin heavy chain composition in human skeletal muscle. Med. Sci. Sports Exerc. 43, 1619–1625.

Stewart, J., Manmathan, G., Wilkinson, P., 2017. Primary prevention of cardiovascular disease: a review of contemporary guidance and literature. JRSM Cardiovasc. Dis. 6. 2048004016687211.

Uth, N., Sorensen, H., Overgaard, K., Pedersen, P.K., 2004. Estimation of VO2max from the ratio between HRmax and HRrest—the heart rate ratio method. Eur. J. Appl. Physiol. 91, 111–115.

Wagner, H., Orwat, M., Hinz, M., Pfusterschmied, J., Bacharach, D.W., von Duvillard, S.P., Muller, E., 2016. Testing game-based performance in team-handball. J. Strength Cond. Res. 30, 2794–2801.

Wanaka, A., Malbon, C.C., Matsumoto, M., Kamada, T., Tohyama, M., 1989. Presence of catecholaminergic axon-terminals containing beta-adrenergic receptor in the periventricular zone of the rat hypothalamus. Brain Res. 479, 190–193.

Zagatto, A.M., Papoti, M., Gobatto, C.A., 2008. Anaerobic capacity may not be determined by critical power model in elite table tennis players. J. Sports Sci. Med. 7, 54–59.

INDEX

Note: Page numbers followed by *f* indicate figures, and *t* indicate tables.

Made in the USA
Middletown, DE
30 March 2023